DICTIONARY OF SOIL MECHANICS AND FOUNDATION ENGINEERING

DICTIONARY OF SOIL MECHANICS AND FOUNDATION ENGINEERING

John A Barker

BSc Eng (Lond), CEng, MICE, FINucE, MBIM

Construction Press

London and New York

Construction Press, Longman House, Burnt Mill, Harlow, Essex,
United Kingdom

A subsidiary company of Longman Group Ltd, London.
Associated companies, branches and representatives throughout the world.

Published in the United States of America by Longman Inc, New York.

First published 1981

British Library Cataloguing in Publication Data

Barker, John A
Dictionary of soil mechanics and foundation engineering.
1. Foundations – Dictionaries
I. Title
624.1'J'0321 TA775

ISBN 0-86095-885-X

Printed in Great Britain by
William Clowes (Beccles) Ltd, Beccles and London.

PREFACE

This dictionary of Soil Mechanics and Foundation Engineering brings together, for the first time ever, the exact meanings of all terms used in these subjects. It contains over 2500 entries giving definitions and supplementary information on the terms and apparatus used in the following major sub-divisions:—

Borehole equipment
Electronic testing equipment
Excavation plant
Ground exploration equipment
Ground water conditions
Ground workers' tools
In-situ instrumentation
Properties of soil
Shoring equipment
Site investigation equipment
Soil mechanics testing equipment
Theory of soil mechanics
Theory of foundation engineering
Types of foundation
Types of retaining structures
Types of soil

Thus this dictionary is a vital new tool for everyone working in, or studying, architecture, building, construction, civil engineering, quantity surveying, utility services etc. It is especially needed by persons studying or working in the fields of civil engineering design and construction, design of structures, foundation engineering, geology, geotechnics, ground engineering, ground exploration, mineral prospecting, mining, offshore engineering and surveying.

To keep the cost of the dictionary as low as possible no diagrams have been included. Where a diagram would have been expected, the dictionary entry has been greatly extended to give a detailed description of the apparatus or topic as well as the strict definition.

The dictionary is up to date, with entries such as Geotextile and SUSIE-PHONE, which was introduced into the subject language in 1980. Other entries, such as Skin friction, have been brought up to date to allow for new divisions of the term, such as Adhesion factor.

Major terms in the subject abroad, especially the United States of America, have also been included and much thought has been given to the cross-referencing of all the entries.

John A Barker
Shoebury, Essex January 1981

A

AASHO
American Association of State Highway Officials.

AASHO classification system
A soil classification system dating from 1928 and mainly used by highway engineers. All soils are divided into seven groups designated A-1 through to A-7 based on their decreasing order of stability. Highly organic soils are designated as A-8.
See also **Group index.**

AASHO rammer
A 4.5 kg rammer, having a 450 mm drop, and used to compact soil for a CBR test.
See also **Compaction rammer.**

Abbot compaction test
The apparatus consists of a metal cylinder 52 mm in internal diameter and 400 mm in effective height, which can be clamped to a metal base. A 2.5 kg rammer with a circular face, 50 mm in diameter, can be lifted up and dropped inside the cylinder through a height of 350 mm above the base. The upper portion of the stem of the rammer is graduated in millimetres.
For the test, the soil is first air dried and pulverized to pass a 4.75 mm sieve. Six or more 200 g samples of air-dried soil are taken and measured quantities of water are added to the various samples to give water contents ranging on either side of the expected optimum moisture value. The height of the compacted sample is determined by reading on the graduated stem of the rammer. The volume of the compacted sample is calculated from the known values of its height and cross-section. The known dry mass of the sample divided by its volume gives the dry density.
See also **Proctor compaction test.**

Absolute surface
The amount of surface of the various soil phases which are in contact with each other.

Absorbing well (or **well drain**)
A well to drain away water.
See also **Vertical sand drain.**

Absorption limit
The point at which the pore water suction, in a fully saturated soil, is equal to zero.

Abutment
The support to a bridge. The anchorage of a suspension bridge, or what the limb of an arch bridge end rests on.
See also **Cantilever, Gravity, Spill through** and **U abutment**, and **Wing wall.**

Abutment wall
See **Wing wall.**

AC system
See **Airfield soil classification.**

Acceptance tests
Specified tests for soil to be used as a fill material in embankments and earth dams to avoid excessive variation of properties.

Access
An approach to a building site.

ACI
American Concrete Institute.

Acidity test
See **pH scale** and **pH meter.**

Acoustic piezometer
The piezometer tip comprises a porous element integral with a diaphragm vibrating wire transducer, installed either in a borehole or by pushing to shallow depths in soft soil. An armoured cable connects the transducer to a remote read-out terminal unit. The mode of operation is by electrical plucking and readings are made by measuring the time for 100 cycles and converting the time into a frequency and hence pressure.

Acoustic piezometer tip
See under **Piezometer tips.**

Active earth pressure (p_a)
The horizontal pressure of the soil against the back of a retaining wall resulting from a very slight movement of the wall away from the soil.
See also **Coefficient of active earth pressure** and **Passive earth pressure.**

Active layer
The layer at the surface of the ground which moves seasonally owing to volume changes, expanding when frozen in winter, and shrinking when it thaws and dries in summer.

Activity (A)
This measures the behaviour of the clay mineral fraction only of a soil and is the ratio of the plasticity index to the clay fraction (in percentages) of the soil.

$$A = \frac{PI\%}{\%\text{ by mass of clay sized particles}}$$

Adfreezing strength
The frozen soil shear strength at the surface of adfreeze to the structure foundation material.

Adhesion factor
A dimensionless factor relating the shear strength of remoulded soil to that of the undrained cohesion of the soil. This factor would be needed to estimate the peripheral resistance of a driven pile or caisson.

Adit
An opening or passage, especially into a mine or tunnel.

Adjoining owner
A person or persons owning lands, a building or portion of a building which abuts upon or is within 3 m of the proposed building operation. Where the excavations for a new building are below the foundations of an adjoining building and, particularly where underpinning has to be carried out, buildings up to a distance of 6 m should be included.

Admissible load
See **Safe bearing capacity**.

Adobe
A term used in semi-arid regions to describe light coloured soils ranging from very plastic clays to sandy silts.

Adsorbed layer
See **Adsorbed water**.

Adsorbed water
Capillary water in the form of a very thin layer, average thickness 0.005 μ, surrounding an individual soil particle.
See also **Capillary water**.

Aeolian
An adjective applied to soil deposits formed of windblown sands and silts, e.g. sand dunes, loess and brick-earth.

Aeration zone
The depth of soil from the ground surface level to the ground-water level. It can be considered to be split into three layers: an upper soil belt, an intermediate belt and a lower capillary fringe.
See also **Capillary water, Held water** and **Topsoil**.

Affine strain
If similar portions of a homogeneous body remain similar after deformation the strain is defined as affine or homogeneous.

Aggregate
Cleaned natural soil used for making concrete.
Fine aggregate is sand, crushed gravel or crushed stone less than 5 mm maximum size.
Coarse aggregate is natural gravel, crushed gravel or crushed stone greater than 5 mm minimum size.

Aggregation
See **Flocculated structure**.

Agricultural drain (or **field drain**.)
Unsocketed, unglazed, earthenware, plastic or porous concrete pipes usually about 75 mm internal diameter, laid end to end without closing the joints so as to drain the subsoil.
See also **French drain**.

Air content (a_c)
The ratio of the volume of voids of a soil filled with air to the total volume of voids expressed as a percentage.

Air-dry soil
A term which designates the moisture condition that approaches equilibrium with an atmosphere that is unsaturated with water. Consequently, depending upon the degree of saturation and the temperature, the water in an air-dry soil may be under tension between pF 4.5 and 7.

Airfield soil classification
A soil classification system published by Arthur Casagrande in the USA in 1948. Each soil is allotted two letters: a prefix depending on the predominate particle size, and a suffix related to the engineering properties (consistency limits).

Air lock
A chamber used in compressed-air work, having one door to the open air and another to the space under compressed air. It may be filled with compressed air or reduced to atmospheric pressure to enable men and/or materials to be passed to or from the compressed-air space, while maintaining pressure in the latter.

Air void ratio
The ratio of the volume of air present in a soil to the total volume of the soil.

Air voids line
A line showing the dry density/moisture content relationship for soil containing a constant percentage of air voids. The line can be calculated from the equation.

$$\rho_d = \rho_w \frac{1 - \frac{V_a}{100}}{\frac{1}{G_s} + \frac{w}{100}}$$

where ρ_d is the dry density of the soil,
ρ_w is the density of water,
V_a is the volume of air voids in the soil expressed as a percentage of the total volume of the soil,
G_s is the specific gravity of the soil particles,
and w is the moisture content expressed as a percentage of the mass of dry soil.

Air-water cylinder
See under **Pressure system**.

Alec Skempton
See **Skempton, Alec**.

A–line
See under **Plasticity chart**.

Allan Hazen
See **Hazen, Allan**.

Allowable bearing capacity
See **Safe bearing capacity**.

Allowable bearing pressure
The maximum allowable net loading intensity at the base of a foundation, taking into account the ultimate bearing capacity, the amount and kind of settlement expected and the ability of the given structure to take up this settlement. It is, therefore, a combined function of both the site conditions (including all construction in the vicinity) and the characteristics of the particular structure it is proposed to erect.

Allowable load
The load which may be safely applied to a pile after taking into account its ultimate bearing capacity, negative friction, pile spacing, overall bearing capacity of the ground below the piles and allowable settlement.

Allowable settlement
That settlement for a particular structure which is limited by (1) allowable total settlement and (2) allowable differential settlement, both for the structure as a whole and between parts of the structure.

Alluvial fan
Soil transported by streams of flowing water which, when the gradient of the stream decreases abruptly resulting in a sudden decrease in its velocity, the soil material deposits to form an alluvial fan.

Alluvial soil
See **Alluvium**.

Alluvium
A term applied mainly to a river-laid gravel, sand, silt or clay and filling a valley bottom and forming the flood plain. Owing to the mode of formation peat beds may occur interbedded with the detrital material. The word is often loosely applied to other fluviatile deposits such as terraces of gravel on valley slopes above the flood plain level, delta deposits and marine terraces.

Aluminium foil sampler
See under **Delft sampler**.

AM-9
A registered chemical grouting product of the American Cyanamid Company. It is a mixture of two organic compounds, acrylamide and methylene bisacrylamide, available as a water soluble powder.

Ambient temperature
The temperature of the surrounding air.

Analogue indicating system
A system which monitors performance of soil tests by using normal 'meter type' calibrated indicators.
See also **Signal conditioning module and 'Dot' recording system**.

Anchorage
See **Anchor block** and **Anchor pile**.

Anchorage

Anchorage distance
The distance behind a wall at which the anchor block must be placed so as to ensure that it will not slip with the tied wall, and that it anchors it effectively.

Anchor block
A wall, pile, plate, beam or massive block buried in a soil to provide an anchorage for tie rods, guy wires or suspension bridge cables.
See also **Ground anchor**.

Anchored bulkhead
This consists of vertical interlocking sheet piles which are embedded in the soil sufficiently deeply to obtain a certain degree of support below a dredge line and which are restrained near the top by tie rods connected to some form of anchor.

Anchored wall
A wall that is fixed at its base but supported against overturning by tie rods or struts near its top.

Anchor pile
See **Tension pile**.

Andreasen pipette
A 25 ml capacity pipette conforming to British standards and used in a soil sedimentation test.

Angledozer
A blade similar to the bulldozer but which can be angled horizontally about 25 degrees to either side and can be tilted about its centre to a maximum of 300 mm from the horizontal at each end.
See also **Back ripper**, **Bulldozer** and **Tiltdozer**.

Angled pile
See **Raker pile**.

Angle of contact
See **Contact angle**.

Angle of internal friction (ϕ_e)
The angle ϕ_e in the original Coulomb equation adapted by Hvorslev in 1937:

$$\tau_f = C_e + \sigma_n' \tan \phi_e$$

where τ_f is the soil shear strength,
C_e is the true cohesion and is not a constant but is a function of the soil water content,
and σ_n' is the effective normal stress.

Angle of obliquity
The angle that a reaction on a plane of sliding makes with the normal to that plane. At the point of sliding this angle is known as the angle of friction.

Angle of repose
For any given dry granular material, the steepest angle to the horizontal at which a heap of it will stand up.

Angle of shearing resistance (ϕ)
The angle ϕ in Coulomb's original equation

$$\tau = c + \sigma_n \tan \phi$$

where τ is the soil shear strength,
c is the apparent cohesion,
and σ_n is the total normal stress.
For completely dry, or submerged soils without cohesion, such as clean sands, the angle of shearing resistance is approximately the angle of repose.
See also **Envelope of failure** and **Angle of internal friction.**

Angle of wall friction
The angle between the direction of the resultant active earth pressure on the back of a retaining wall, and the normal to the back of the wall.
The angle is considered positive when the resultant reaction is orientated such that its tangential component acts in an upward direction.

Angularity factor
The ratio S_s/S_i where S_s is the specific surface of soil particles and S_i is the specific surface of spheres of the same sieve size.

Angular particles
A description of the angular shape of some course soil particles.
See also **Rounded, Subangular** and **Subrounded particles.**

Anion
A negatively charged atom.
See also **Cation** and **Co-ions.**

Anisotropic (or **anisotropy**)
Having different coefficients in different directions.
Since most soils are deposited horizontally they thus have different coefficients of permeability and different shear strengths, horizontally and vertically.

Anode
See under **Electro-osmosis.**

Anti-heave measures
Precautions taken to prevent soil heave.

Anvil
The part of a power-operated hammer which receives the blow of the ram and transmits it to the pile being driven into the ground.

A-parameter
See under **Pore pressure parameters.**

Apparent cohesion (c)
The value of c in the original Coulomb formula

$$\tau = c + \sigma_n (\tan \phi)$$

where τ is the shear strength of the soil,
σ_n is the total normal stress on the failure surface
and ϕ is the angle of shearing resistance.
See also **True cohesion.**

Apparent velocity (v)
The average rate of flow of a liquid (usually water) across a unit area of soil. See also **Darcy's law.**

Approach pit
A narrow pit dug to a suitable depth and as small a distance as possible below and adjacent to that part of the foundation of a structure which is to be underpinned.

Aquaducer
A portable, earth settlement measuring instrument.

Aquagel
A trade name for dry powdered bentonite.

Aquicludes
Impervious strata covering both the top and bottom of confined water.

Aquifier
A permeable water-bearing soil stratum sandwiched between impermeable strata above and below it.
See also **Artesian water.**

Arching
A phenomenon of cohesive soil which enables it to transfer its pressure from the back of a 'timber', that has been removed from the side of a trench, to the back of the remaining adjacent timbers.

AREA
American Railway Engineering Association.

Area ratio (A_r)
The degree of disturbance of a soil sample, due to the wall thickness of the cutting edge of the sampling tube, and the manner in which it is forced into the in situ soil:

$$Ar = \frac{D_e^2 - D_i^2}{D_e^2} \times 100\%$$

where D_e is the external diameter of the cutting edge
and D_i is the internal diameter of the sampling tube.
See also **Inside clearance ratio.**

Armco pile
A thin-walled concrete tube pile having a concrete plug at its lower end. It is driven into the ground by means of a driving core and head. After the pile is driven to the correct depth, the driving core is removed and the internal void filled with concrete.

Artesian water
Ground-water confined in a permeable stratum beneath an impervious stratum and under sufficient pressure to raise it above the base of the confining bed when the latter is penetrated by a well.
Such wells may overflow at the surface (flowing artesian) or the artesian water may fail to reach the surface (non-flowing or sub-artesian). The term should not be applied to wells, however deep, if the water does not rise above the point at which it was struck.

Artesian well
See under **Well**.

Arthur Casagrande
See **Casagrande, Arthur**.

ASCE
American Society of Civil Engineers.

Aseismic region
A region not liable to earthquakes.

ASTM
American Society for Testing Materials.

Atomic bonds
The two main bonds are the primary or high-energy bonds which hold atoms together and the secondary or weak bonds which link molecules.
See under **Ionic bond** (primary bond) and **Van der Waal forces** (secondary bond).

Attapulgite clay
Clay in which attapulgite is the predominant clay mineral. It has similar properties to a bentonite clay.

Atterburg limits
A series of empirical tests for fine particled clay soils which measure the water contents at which certain changes in the physical behaviour of the soil can be seen.
These tests are the liquid limit, plastic limit, shrinkage limit, liquidity index, and plasticity index, from which it is possible to estimate the engineering properties of the soil.

Auger
An instrument for boring holes.
In soil exploration augers can be either hand or mechanically operated. They are used for bringing up disturbed soil samples from a borehole, or for advancing a borehole down to the point where an undisturbed soil sample is to be taken. For this, the auger is removed and a sharp-edged soil sampler is lowered into the borehole and forced into the soil.
See also the following types of auger – **Bucket, Buda, Cup, Disc, Dutch, Gravel, Helical, Hollow stem, Iwan, Posthole, Scotch eyed, Shall, Spoon, Split barrel, Vicksberg** and **Worm**.

Auger board
A slotted steel plate placed at the top of a borehole to provide a platform on which to suspend the auger drilling rods when making and breaking joints.

Auger pile
See **Bored pile**.

Auger sample
See under **Sample**.

Auger vane
See under **Vane test**.

Autographic oedometer
A soil consolidation test machine which automatically records the soil compression on a constantly revolving drum chart.

Autographic unconfined compression apparatus
A portable apparatus, originally designed by the Building Research Establishment, to determine the unconfined shear strength of a cohesive soil sample at site.
The lightweight apparatus consists of two platens between which the soil sample is placed and the load applied by means of a hand operated leadscrew which extends a calibrated helical spring. A moving arm, complete with pencil, pivots about the lower platen and draws a stress/strain curve on the chart fitted to the front of the apparatus, as the load is applied. Strength values can be determined from the drawn chart curve by using a specially calibrated transparent mask over the chart.
See also **Unconfined compression test**.

Automatic compaction machine
A machine to automatically compact soil specimens, usually for Proctor and CBR tests.

Auxiliary drains (also known as **Chevron, Herringbone** or **Vee drains**)
Drains provided between main butress drains in the form of narrow stone filled trenches, tending up a slope from points in the main buttress drains at an angle of 45°, and meeting midway between the main drains.
They are usually 750 mm deep by 450 mm wide and spaced from 3 to 6 metres apart. Their function is to intercept water flowing down the face of a slope and to add to the stability of the surface of the slope.

B

Back acter (or **Drag shovel** or **Hoe**)
A bucket type shovel which excavates radially backwards and downwards from the machine to which it is mounted.

Back drain
A perforated drain pipe having a minimum internal diameter of 100 mm, placed longitudinally behind the bottom of a retaining wall and surrounded by a stone filter, to drain the backfill.
This reduces the water pressure acting on the rear of the wall.

Backfill
The placing of fill material (soil) in confined spaces, such as around constructed foundations.

Back hoe
See **Back acter**.

Back pressure
A pressure used in soil testing to ensure saturation of the soil sample.

Back prop
A raking strut used to transfer the mass of 'timber' to the ground in deep trenches. It is usually placed below every second or third frame.

Back ripper
Short powerful spikes pivoted on the back of the blade of an angledozer, bulldozer or tiltdozer and mounted to face backwards. They bite into the hard soil when the dozer moves backwards and float on the surface when the dozer moves forward.

Back scatter density
See under **Nuclear densometer**.

Bailer
See **Sand pump**.

Ball clay
A clay containing illite and sometimes montmorillonite as well as kaolinite. Mainly used for the manufacture of pottery, floor and wall tiles.

Ball cone clamp
A simple device for gripping the borehole drilling rods during extracting operations. The device consists of one or two rows of hardened steel balls set in a housing. When a load is applied, the balls are pressed against the drill rods, producing a positive grip.

Balloon densometer
An apparatus to determine the in situ density of compacted soil, A plastic balloon is expanded by water pressure in an excavated hole and the water used to fill the balloon is measured on a graduated cylinder.
See also **Sand pouring cylinder**.

Banksman
A person who stands near the edge of a deep excavation to direct, by hand actions, the driver of a crane or excavator so that he can correctly position the excavating bucket or grab within the excavation.
See also **Signal man** and **Spotter**.

Barco rammer
See **Power rammer**.

Bar dog
A long handled wrench used for making and breaking boring tool rod joints.
See also **Hand dog**.

Barentsen hand sounding apparatus
A static cone penetrometer.
It is used by pressing sounding tubes, pressure tubes and a cone vertically into the soil by one or two people using a special pushing yoke. The cone resistance is measured by pressing on the rods only using the driving appliance which has a pressure dial gauge attached. Depths up to 10 m or resistances up to 0.15 kN/m^2 can be obtained.

Barium sulphate test
A quick chemical colour test to approximately determine the pH value of soil.

Barrettes
Large oblong piles constructed by diaphragm walling techniques using bentonite to support the sides of the excavation. They are suitable where very heavy loads have to be carried to great depth.

Barrow
A wheelbarrow.

Base course
See under **Pavement**.

Base exchange
A reversible chemical process. A hydrogen clay (one with adsorbed hydrogen ions) can be changed to a sodium clay (with adsorbed sodium ions) by the infiltration of sodium solutions. This decreases the permeability of the clay.

Base material
See under **Pavement**.

Basement wall
This is a normal type of L-shaped cantilever wall, but with the added resistance against overturning of the ground floor slab.

Batter
A slope.

Batter board
See **Profile**.

Batter drain
See **Buttress drain**.

Batter peg
A peg driven into the ground to set out the limits of a slope.

Batter pier
An underground shaft having sloping sides. The slope is usually limited to 1 horizontal to 6 vertical.

Batter pile
See **Raker pile**.

Batter post
See **Batter peg**.

Batter rule
A board with one edge cut to the batter of a wall or slope and fitted with a plumbline and bob.

Beam grillage
See **Crib**.

Bearer
A horizontal timber which supports other timbers.

Bearing capacity
The load per unit area which the ground, or a pile, can carry.

Bearing pile
A pile driven or formed in the ground for transmitting the mass of a structure to the soil by the resistance developed at the pile point or base and by friction along its surface. If it supports the load mainly by the resistance developed at its point or base, it is referred to as an end bearing pile; if mainly by friction along its surface, as a friction pile.
See also under **Foundation** for types of pile.

Bearing stratum
The stratum (or formation or bed) which has been chosen as the most economical or suitable to carry the load in question.

Bearing test
See **Plate bearing test.**

Bedding plane
The plane of junction between adjacent laminae or strata in deposits of sedimentary origin.

Bedrock
Any hard rock-bed underlying soft deposits classed as soil in the engineering sense. Geologists restrict the word to the formation, hard or soft, that underlies drift deposits.

Beeswax
One of the additives to paraffin wax which is used for sealing soil samples that are to be stored.
See also **Carnaubawax**, **Ceresine** and **Paraffin wax.**

Beidellite
A clay mineral similar to montmorillonite.

Bell box
A recovery tool for boring equipment. It consists of an internally tapered bowl which has a spring-loaded hinged flap at the top. The bowl is fixed between the ends of a long steel two-pronged fork. The top of the fork is fixed to the bottom of a string of boring rods.
In operation the internal tapering bowl guides the 'lost' tool through the hinged flap which then closes behind the collar of the 'lost' tool.

Belled out
See **Under-reaming.**

Belled pier
See **Under-reaming.**

Belling bucket
A special bucket consisting of a cylinder with two cutting blades, pivoted at the top, that fold inside the cylinder while the bucket is lowered down a hole. The bucket replaces the auger to under-ream the bottom of an augered hole.

Bellofram
A rolling diaphragm used in conjunction with ball bushings to provide a watertight seal and frictionless movement of a load shaft moving in and out of a triaxial compression cell.

Bell

Bell pits
Medieval bell shaped pits dug down to exploit clay – ironstones where the drift cover was thin.
This type of working was commonly used until the seventeenth century. In general the shafts rarely exceeded 12 m in depth and their diameter was usually about 1.2 m. Radial mining from the bottom of the shaft continued until the area of extraction was such that natural or artificial support was not feasible.

Bell screw
A steel cone, threaded along the inside of the cone. It is fixed to the bottom of boring rods and used for the recovery of 'lost' boring tools by screwing onto the broken ends.

Bench
See **Berm**.

Benched foundation
A foundation on a sloping bearing stratum, cut in steps to ensure that it will not slide when concreted and loaded up.

Benching
(1) A berm above a ditch.
(2) Forming horizontal steps in a foundation on sloping ground.

Bench mark
An Ordnance Survey reference point from which to commence a levelling survey.

Bends
A slang name for Caisson disease.

Bentonite
A clay largely comprisng the mineral sodium montmorillonite.
When in contact with water it has a characteristic ability to swell to many times its dry volume, resulting in suspensions of jelly-like consistency which when agitated become fluid but revert to the jelly form if allowed to stand. This phenomenon, known as thixotropy, is one of the four main properties of bentonite/water suspensions which are utilised in civil engineering practice. The other three properties are lubricity, plastering and sealing ability.
See also **Attapulgite clay**.

Bentonite/cement pellets
These consist of a mixture of three parts bentonite clay and one part cement by volume. They provide a simple method of backfilling boreholes with grout without the need for grout mixers or pumps.

Bentonite pellets
Pellets made of bentonite. These are used to drop through the water in a bore-hole, where they swell to form a low permeability plug such as is required above a piezometer tip or well point.

Berkbent
A trade name for bentonite.

Berm
A horizontal ledge at the top, or into the side, of an embankment or cutting.

Bernoulli's equation
See under **Hydraulic head.**

BGS
British Geological Society.

Biat
A 'timber' bearer giving support to guard rails, decking, walkways, etc.

Bind
See under **Mudstone.**

Binder
The clay or silt in hoggin.

Bingham model
A complex composite rheological model to represent an elastoplastic response and a delayed viscoelastic response of soil.
See also **Rheological model.**

Biot's theory
A theory of three dimensional consolidation developed directly from the elastic theory for a homogeneous isotropic fully saturated soil and first published by Biot in 1941.

Bishop piezometer tip
See under **Piezometer tip.**

Bishop sand sampler
See **Sand pump.**

Bit
See **Chopping bit.**

Bitch
(1) A fastening of iron or steel used for securing heavy excavation timbers which cross each other. It is similar to a dog but has one of its ends at right angles to the other.
(2) A flat forked rod which slides under a drilling rod joint to hold the rod suspended through the auger board when making and breaking the drilling rod joints.

Bitumen
A viscous liquid, or a solid, consisting essentially of hydrocarbons and their derivatives. It softens gradually when heated. It is black or brown in colour and is used for waterproofing.

Bitumen grout
See under **Grout.**

Bituminous surface
See under **Pavement.**

Black cotton soil
Black, or sometimes brown, soil which shows very 'quick' marked volume changes with change in moisture content.

Blade
A vertically curved steel plate (blade) positioned at a fixed distance in front of a bulldozer or angledozer by arms, which are, in turn, anchored at pivots along each side of the bulldozer or angledozer tractor.

Blade grader
See **Grader**.

Blading back
Pushing, with a grader, soil in a windrow back to the original position.

Blanket
See **Drainage blanket**.

Blind drain
See **Rubble drain**.

Blinding
A thin layer of weak concrete, or other suitable material, placed directly on the soil beneath a foundation.

B-line
See under **Plasticity chart**.

Bloated clay
Expanded clay.

Blob foundation
A term used in the building industry for an isolated shallow footing or base.

Block diagram
See **skeletal diagram**.

Blocking
A 'timber' block used as a distance piece or packing, eg between a waling and the temporary lining of an excavation, to permit the insertion and erection of vertical reinforcement in retaining walls or other permanent construction.

Block pavement
A road wearing surface made of blocks of stone or wood.

Block plan
A line drawing of the plan of a building and its surroundings drawn to a small scale. It may also show the plan positions of the boreholes.

Block sample
See **Hand carved sample**.

Blow
See **Boil**.

Blow count
The number of blows of a pile driving hammer required to penetrate the pile into the ground for a specified distance.
See also **Rate of penetration**.

Blown sand
Sand deposited after transport by wind, eg a dune sand.

Blow up
A compression failure in which buckling of a pavement slab occurs.

Blue test
See **Methylene blue test**.

BM
Bench mark.

Bog
Soft waterlogged ground composed largely of peat or mud.

Boglime
A fine-particled white powdery calcareous deposit precipitated by plants in ponds. Commonly associated with peat.

Boil
A flow of water and soil, usually fine sand or silt, into the bottom of an excavation, due to the water pressure outside the excavation being greater than that inside. It starts as a small spring and like piping may increase rapidly.

Bolt
See **Tie rod**.

Boning rod
A T-shaped instrument for obtaining levels and falls, usually made from timber. The upright is one metre or more in length and the crosshead about 200 mm long.

Book clay
See under **Varved clay**.

Bootstrap test
Test loading a pile by jacking against the reaction provided by a beam spanning between two anchor piles driven on opposite sides of the test pile.

Bored pier
See **Drilled pier**.

Bored pile
A pile formed by pouring concrete into a hole formed in the ground. Usually contains reinforcement.

Borehole
A hole driven into the ground to get information about the strata, or to release water pressure by vertical sand drains, or to obtain water, oil, gas, salt or sulphur.

Borehole camera
A television or film camera that may be lowered into a borehole. The strata can be observed on a television screen or recorded for later viewing on a cylindrical screen where it may be examined as if the viewer were in the borehole.

Borehole extensometer
See **Magnetic probe** and **Rod extensometer**.

Borehole log
A document showing a complete record of a soil boring.
The log is initially prepared at site by the Engineer or foreman in charge of the boring operation. After all the laboratory soil tests have been analysed a more detailed log is prepared which is included in the final site investigation report presented to the clients.
The borehole log should show the following information: —
(1) Name of the client, name of the person in charge of the boring team, location, details of present weather and previous few days weather if known, borehole and job reference number, date boring commenced and finished.
(2) Method of augering, diameter of borehole, type of casing used, existing ground level.
(3) Description of each soil encountered, level of each change in soil.
(4) Level of water first encountered, times of changes in water level, water seepage level.
(5) Serial number, depth and type of each sample taken.
(6) Reference number, depth and type of each test taken inside the borehole.
(7) Level at which boring stopped.

Borehole periscope
A periscope consisting of a brass tube with a telescope at its upper end and a light source and a mirror at the lower end, used for examination of dry boreholes up to a depth of 25 metres.

Borehole reference stud
See **Geonor settlement probe** and **Heave gauge**.

Borehole sample
See under **Sample**.

Boring
Obtaining a knowledge of the soil strata at various depths by boring holes, usually 100 mm in diameter.
See also **Auger**.

Boring rig
See **Rig**.

Boring rod
A square shanked steel rod having cylindrical sections at each end; one end has a threaded stud and the other end is tapped. They are usually made in standard lengths of 0.5, 1.5 and 3 metres.
See also **Swivel rod**.

Borros point
See under **Geonor settlement probe**.

Borrow pit
A place from which to excavate soil for use elsewhere.

Bottle roller
An apparatus for rotating glass jars about their longitudinal axis as required in a soil sedimentation test.

Boulder clay
A deposit of clay or sandy clay of glacial origin containing stones of various sizes scattered irregularly throughout its mass. The stones are not necessarily all of 'boulder' size.
See also **Raisin cake structure** and **Till**.

Boulder extractor
A special auger head consisting of a single three-quarter helical screw at its base which can underbore and retrieve a boulder from a borehole. This extractor can be used only if the boulder is not more than one third the diameter of the borehole. If the boulder is too large or too tightly held to be picked up by the extractor then it must first be broken up or loosened.

Boulder gravel
See **Gravel**.

Boulders
Subangular to rounded rock fragments, the lower limit of size being about 200 mm in diameter.

Boundary piezometer
A disc piezometer used to measure pore water pressure at the interface between a structure and the soil.

Boundary pressure cell
Any hydraulic, pneumatic or electrically operated pressure cell which has been designed for use on a boundary, ie it has at least one flat surface.

Boussinesq theory
A mathematical stress distribution theory in a semi-infinite elastic body caused by a point load applied to the surface of the body, after Boussinesq.
The intensities of the stresses at a point O in the ground and defined by coordinates x, y and z from the surface point load P are as follows:

Vertical direct stress on horizontal plane at depth z

$$= \frac{3Pz^3}{2\pi R^5}$$

where R is the diagonal distance from the point load P to point O.

Vertical shear stress

$$= \frac{3Prz^2}{2\pi R^5}$$

where r is the projected horizontal distance from the point load P to point O.

Bowl

Horizontal direct shear (parallel to the y axis)

$$= \frac{P}{2z}\left[\frac{3y^2z}{R^5} - (1-2\mu)\left(\frac{y^2-x^2}{Rr^2(R+z)} + \frac{x^2z}{R^3r^2}\right)\right]$$

where μ = Poisson's ratio.

Bowl mixer
See **Mixer**.

Bowl scraper
See **Scraper**.

Box dam
A rectangular-shaped cofferdam.

Box heading
A heading in which the walls and roof are close 'timbered'.

Box pile
A pile made of rolled steel sections welded together to form a hollow pile.

B-parameter
See under **Pore pressure parameters**.

BPR
American Bureau of Public Roads.

Brace
See **Bracing**.

Bracing
The internal system of walings, struts, and other members which enable an excavation to resist external pressures.

Braided stream
In its lower reaches, a stream overloaded with sediments deposits the material in its bed and the bed of the stream is raised. The water then spills over into an adjacent depression and establishes a new channel. Such a stream is said to be braided because it consists of a series of rivulets continually joining each other and then separating.

Brass piezometer tip
See under **Piezometer tips**.

BRE
Building Research Establishment.

Breaker tools
Large steel chisels, etc, which fit into the end of a pneumatic tool for the purpose of breaking up soil and other materials.

Break in failure
A type of foundation failure where the soil beneath the foundation is uniformily compressed and bulges uniformily around the outside of the foundation.
See also **Rotational failure** and **Rotational slide**.

Breccia
A breccia consists of angular rock fragments cemented together in a matrix of finer material.
A breccia may occur as a filling in fault fissures or on lines of intense differential earth movement or as a stratigraphical unit, where the fragments are accumulations of talus formed usually, under desert conditions.
A breccia is distinguishable from a conglomerate in that its rock fragments show no sign of abrasion due to transportation by water.

Brick clay
See **Brick-earth**.

Brick-earth
A name given to any natural material suitable for making bricks, but particularly to buff or biscuit-coloured homogeneous structureless deposits of silt or finer material occurring in spreads and patches along the slopes of the Thames Estuary and elsewhere in South-eastern England.

Bridge abutment
See **Abutment**.

Bridge circuit
See **Wheatstone bridge circuit**.

Bridge foundation
A foundation to support part of a bridge.
See also **Abutment** and **Pier**.

Bridge pier
See **Pier**.

Briefly frozen soil
Soil that has been frozen from a few hours to a few days.
See also **Frozen soil**.

British soil classification
A soil classification system that can either be used for a rapid assessment, or a complete laboratory procedure, of the soil type.

Brittle failure
A sharp compression failure of a soil sample, in an unconfined or triaxial compression test, which clearly expresses shear failure.
See also **Intermediate** and **Plastic failure**.

Brob
A fastening of iron with its head bent at right angles to the shaft and used for connecting 'timbers'.

Brooming
The breaking of a timber pile during driving.
If a blow-by-blow driving log is being kept, brooming will usually be detected by an increase in the rate of penetration.

BRS
Building Research Station.

Brucite
One of the magnesium clay minerals $Mg_3 (OH)_6$ having a plate form structure.

Brunspile
A prestressed sectional pile where the connections are made with a steel ferrule being forced into a slightly tapered connector.

BS
British Standard such as BS 1377 on Standard soil tests and BS 1924 on Methods of test for stabilized soils.
See also **CP**.

BSCS
British Soil Classification System.

BSI
British Standards Institution.

Bucket auger
An auger with a large cylindrical bucket immediately above it to collect the augered soil.

Bucket scraper
See **Scraper bucket**.

Buda auger
A continuous type of worm or helical auger, after Buda.

Building inspector
A person employed by a local authority to inspect all aspects of building construction. This would include inspecting the depth of excavation and the condition of the soil at the bottom of trenches etc, before pouring concrete.

Building line
A line fixed as a limit to building work near a road.

Building regulations
British government regulations covering all aspects of building work and foundations.

Building Research Digests
Leaflets issued by the Building Research Establishment on the latest developments of a topic.

Buisman's law
The logarithmic relationship for secondary consolidation without seepage resistance is given by:

$$\frac{p}{h_\alpha \, \Delta\sigma} = \alpha_p + \alpha_s \log_{10} t$$

where p denotes the settlement,
h_α the initial height of the stratum or sample,
$\Delta\sigma$ the pressure increment,
t the time,
and α_p and α_s the constants given by the graph relating the observed settlements to the logarithm of time.

Bulb end pile
See **Franki pile**.

Bulb of pressure
The mass of stressed soil below a loaded foundation.
The term is also used to describe the bulb-shaped lines of equal vertical stress below a footing, derived from the work of Boussinesq.

Bulk density (γ)
The mass of that material per unit volume. It includes the effect of voids whether filled with gas (air) or liquid (water), and is not to be confused with the specific gravity of the individual particles of which the material is composed.
See also **Density**.

Bulkhead
A sheet pile wall which can be 'free' standing but is more usually anchored back near to the top.
See also **Dredge bulkhead** and **Fill bulkhead**.

Bulking
(1) The increase in the volume of excavated material above the volume of the excavation from which it came, often about 50%.
(2) The increase in volume of dry sand when its moisture content increases. This may amount to 40% when 5% of water is added. This increase disappears entirely when the water content is raised to 20%.

Bulk modulus
Modulus of compressibility.

$$\text{Bulk modulus} = \text{Original volume} \times \frac{\text{Change in pressure}}{\text{Change in volume}}$$

Bulldozer
A blade consisting of a bowl which is fixed at right angles to the machine tracks. Its primary function is to retain spoil it excavates and push it in front to a spoil heap. The blade can be raised, lowered or tilted a few centimeters.
See also **Angledozer**, **Back ripper** and **Tiltdozer**.

Bulldozing
Using a bulldozer.

Bull-nosed piezometer tip
See under **Piezometer tips**.

Bull's liver
A term often used by American construction workers for quick silts.

Bungum
A local name for recent alluvial silt, used for such particularly in the London area. Generally applied to soft clay with or without an admixture of silt.

Buoyancy pile
See **Tension pile**.

Buoyant density
See **Submerged density**.

Buoyant foundation
A reinforced concrete raft foundation so designed that its total mass, including all superload, is approximately equal to the mass of the soil or water displaced. It is often used in river estuaries where no foundation is available except silt or mud.

BuRec
American Bureau of Reclamation.

Burger model
A complex composite rheological model to represent a complex viscoelastic response eqivalent to the consolidation response of soil.
See also **Rheological model**.

Burr scraper
A small hand held hardened steel blade with a notch cut into the end.
This notch is pressed round the rim of a sampling tube to remove any burrs so that the enclosed soil sample is not damaged by burrs during extrusion.

Butment
See **Abutment**.

Buttress drain
A stone or rock filled trench drain which is constructed in steps down the surface of a soil slope. Its function is to assist in removing water from a soil slope and to provide weighty buttresses capable of resisting any local tendency to slip.
On slopes of, say, about 2½ : 1 the length of the benches might be 2.5 meters and the height of each step 1 metre with a maximum depth of excavation of 1.75 metres. The width of the trench will usually be between 1 and 2 metres and the drain is usually constructed with a wider trench at the bottom of the slope than at the top.
See also **Auxiliary drains**.

Buttress wall
A vertical retaining wall supported by buttresses monolithic with the front, or back, of the wall and the base slab.

Byatt
See **Biat**.

C

CAA
American Civil Aeronautics Administration.

Cable tool drilling
See **Percussion drilling**.

Cage
See **Pile cage**.

Caisson
A large hollow open ended structure which is sunk through ground or water for the purpose of excavating and placing work at the prescribed depth and which

subsequently becomes an integral part of the permanent work.
See also the following headings:— **Adhesion factor, Box dam, Caisson bell, Caisson disease, Chicago caisson, Cofferdam, Compressed air caisson, Drilled caisson, Drilled pier, Gow caisson, Kentledge, Open caisson** and **Sand island caisson**.

Caisson adhesion
See under **Adhesion factor**.

Caisson bell
The under-reamed portion of a drilled caisson for the purpose of reducing the soil pressure to a desired bearing value. The under-reamed portion is generally known as a bell because of its shape.
The ideal bell is in the shape of a frustum with a vertical side at the bottom. The vertical side may be 150 to 300 mm high. The sloping side or roof of the bell should be at an angle of at least 60 degrees with the horizontal so that the soil will not cave in.

Caisson box
A caisson which is closed at the bottom but open to the atmosphere at the top.

Caisson disease
A disease which can affect workers in compressed air who come out of the air lock too quickly.
It is caused by bubbles of nitrogen coming out of the blood. The only treatment is to take the worker to a special medical lock immediately for recompression and slow decompression. It is commonly called the bends or screws.

Calcareous clay
Marl.

Calcareous soil
A soil containing calcium carbonate.

Calcium carbonate ($CaCO_3$)
The chemical name for chalk or limestone.

Calf dozer
A small bulldozer.

Calgon
See **Sodium hexa meta phosphate**.

Caliche
A soil conglomerate of gravel, sand and clay, the particles of which are cemented by calcium carbonate. Usually formed in semi-arid climates.

California Bearing Ratio (CBR)
A standardised testing procedure begun by the California State Highways Department for comparing the strengths of base courses of pavements (roads) or airstrips.
The site test consists of applying a load to a circular area of soil by means of a stiff plate to which the load is applied by hydraulic jack. The loading plates vary in size from a diameter of 150 mm to 750 mm in steps of 150 mm. Each plate is 25 mm thick. Settlement under load and recovery upon release of load are

measured with dial gauges. Enough load is applied to produce a settlement of from 0.25 to 0.50 mm. The load is then released to one-half that required to seat the plate.
There is also an equivalent laboratory test.

Cambridge drive-in tip
See under **Piezometer tips**.

Camera
See **Borehole camera**.

Camkometer
A self-boring pressuremeter jointly designed by the BRE and Cambridge University.
The instrument consists of an 80 mm external diameter self-boring pressuremeter covered by a rubber membrane that incorporates two small cells for measurement of pore water pressure. Once the instrument is at the desired depth, the membrane expands and transducers enable the pressure response to be converted to electrical impulses, so that an effective stress/strain curve can be plotted and the *in situ* properties of the undisturbed soil derived.

Campshedding
The name given to sheet piling that is placed round a barge quay or jetty.

Cantilever abutment
A cantilever wall that supports the end of a bridge deck above it and backfilled soil behind it.

Cantilever cofferdam
See under **Cofferdam**.

Cantilever footing
A particular type of combined footing where the individual column footings are joined by a beam and the footing under the boundary or external column is made slightly larger. The footing under the internal column is the normal size for the given soil conditions.

Cantilever wall
A reinforced concrete retaining wall consisting of a thin stem and base slab.

Cantledge
See **Kentledge**

Cap
In excavating, a piece of 'timber' placed over the joint where two walings butt to take the thrust of the strut.

Capacity factor
See **Safety factor**.

Cap block
See **Dolly**.

Capillarimeter
An apparatus, such as Beskow's, for determining capillary height of fine textured soils.

Capillarity
The rising of fluid in tubular (or hair-like) spaces above the level of the fluid, in contact with the tubes, in an open vessel.
Capillarity is caused by the surface tension of the fluid and its amount can be roughly estimated at 20°C as follows:

$$\text{capillary rise} = \frac{0.0002}{\text{tube bore in mm}}$$

Capillary fringe zone
See under **Capillary water.**

Capillary pressure
The hydrostatic pressure caused by the curvature of the boundary surface of capillary water.

Capillary saturation
When the voids of a soil above the ground-water level are completely filled with capillary water.

Capillary suction
See **Matrix suction.**

Capillary tension
The tensile stress set up by surface tension forces at the curved surface of a column of capillary water.

Capillary water
Held water, kept above the ground water table by capillary action.
The capillary saturated zone between the water table and the plane of the menisci is called the 'closed capillary fringe'. Above this closed fringe there is an 'open capillary fringe' which reaches to the height of the menisci in the finest pores of the soil. Here the larger pores are not filled with capillary moisture.
See also **Aeration zone.**

Capping piece
A horizontal 'timber' placed over the ends of two walings butted together. It takes the thrust of a strut and transfers it to the walings.

Carbonation
The process of chemical decomposition by which carbon dioxide contained in water combines with the oxides of calcium, iron, magnesium, potassium and sodium. As a result, carbonates or bicarbonates of these metals are formed.

Carlson stress meter
An instrument to measure stress at a boundary. The principle of operation is that a thin film of mercury, when pressurized, causes a diaphragm to deflect and a strain gauge measures the change in output which can be recorded by a suitable 'bridge' circuit.

Carman equation
See **Kozeny-Carman equation.**

Carnaubawax
One of the additives to paraffin wax which is used for sealing soil samples that are to be stored.
See also **Beeswax, Ceresine** and **Paraffin wax.**

Casagrande, Arthur *(born 1902)*
Professor of soil mechanics and foundation engineering at Harvard University. In 1936 he organised the first international conference on soil mechanics and foundation engineering.

Casagrande classification system
See **Airfield soil classification.**

Casagrande drive-in piezometer
See under **Standpipe piezometer.**

Casagrande piezometer tip
See under **Piezometer tips.**

Casagrande plasticity chart
See **Plasticity chart.**

Cased hole
See **Casing** and **Sheeted pit.**

Cased pile
A pile with a permanent shell or casing subsequently filled with concrete.

Casing
Steel tubing used to prevent caving in of the sides of a borehole.

Casing off
This consists of driving an open-ended tube having an internal diameter greater than the pile to be installed, removing the material inside the tube, and then installing the pile through the tube.

Cast in place pile
See **In situ pile.**

Catch drain
See **Grip.**

Catcher
See **Core catcher.**

Catchment
An area of ground draining to a reservoir.

Catena soil
A soil chain where there is a relationship of succession of soil down a slope. For example, the dissolved mineral constituents are mobilised and leached in the topmost strata and are carried down the slope by percolating water to accumulate in the soil below.

Cathode
See under **Electro-osmosis.**

Cathodic protection
A method of protecting below ground pipes and foundations likely to be subjected to corrosion. By reversing the electric current which flows away from a corroding metal, the corrosion process can be arrested.

Cation
A positively charged atom.
See also **Anion** and **Co-ions.**

Cation exchange capacity test
See **Methylene blue test.**

Causeway
A road carried over marsh or water by a soil or rock embankment.

CBR
California Bearing Ratio.

CBR mould
A steel mould 178 mm high with an internal diameter of 152 mm and with external lugs for clamping to the pillars on its baseplate.

Cellular cofferdam
A cofferdam enclosed by a wall consisting of a series of filled cells of circular or other shape in plan.

Cellular raft
A concrete raft in which the intersecting beams form a number of cells.

Cement/bentonite pellets
See **Bentonite/cement pellets.**

Cement bound granular base
See under **Pavement.**

Cement grout
See under **Grout.**

Cement grouting
See **Grouting.**

Cement stabilisation
See **Soil stabilisation.**

Centre of pressure
The point on an area subjected to hydraulic pressure, over which the whole force due to the pressure on the area may be taken to act.

Centric load
An axial load.

Centrifuge
A machine which revolves at very high speed thus producing an overall uniform force on a soil model under test. The model is held in place at the outer end of a revolving arm.

Ceramic piezometer tip
See under **Piezometer tips.**

Ceresine
One of the additives to paraffin wax which is used for sealing soil samples that are to be stored.
See also **Beeswax, Carnaubawax** and **Paraffin wax.**

Chalk
A soft limestone consisting principally of the remains of marine organisms. It has a rigid porous structure and a hardness which varies considerably, depending on the type of chalk. The hardness can be arbitrarily related to the porosity, which in turn can be expressed in terms of the saturation moisture content. For a very hard chalk the saturation moisture content is between 5 and 10%, whilst values of 25 to 30% are found in the case of the softer varieties.

Charles Coulomb
See **Coulomb, Charles.**

Chemical consolidation
A grouting injection process in which the air or water contained in the voids of a granular soil is partially or wholly replaced by chemical fluids which have been designed to gell after injection, thus increasing the strength and reducing the permeability of the soil.

Chemical grout
See under **Grout.**

Chemical grouting
See **Grouting.**

Chemical injection
See **Soil solidification.**

Chemical weathering
The changing of the chemical compounds in a soil due to the action of water and temperature.

Chemise
A non-retaining wall which protects a soil slope against weathering.
See also **Revetment.**

Chevron drains
See **Auxiliary drains.**

Chicago caisson
A caisson best used in clays of at least medium stiffness. The walls are lined with vertical timbers a few millimetres shorter than the distance with which the clay is self supporting. These timbers are then supported laterally by two pairs of steel half rings wedged into position.

Chilled soil
See under **Frosted soil.**

China clay
A clay mainly composed of kaolinite and hence white in colour. Such clay is mainly used for the manufacture of whiteware.

Chisel bit
See **Chopping bit.**

Chisel tool
See **Breaker tools.**

Chock
See **Blocking**.

Chog
See **Blocking**.

Chopping bit
A chisel-shaped bit used in the driving of a borehole. It can have a single chisel face or three spur faces at 120° to each other.

Chunk sample
See **Hand carved sample**.

Churn drilling
See **Percussion drilling**.

CI
Consistency index.

Cill
See **Sill**.

Ciment fondu
High alumina cement. It is noted for its rapid hardening facility.

Circle of stress
See **Mohr's circle of stress**.

Circular arc cofferdam
See under **Cofferdam**.

Circular cell cofferdam
See under **Cofferdam**.

CIRIA
Construction Industry Research and Information Association.

Clack
See **Flapper valve**.

Cladding
A term used for heavy close sheeting in connection with the sinking of shafts.

Clamshell bucket
A bucket used in conjunction with a lift crane for excavating non-cohesive soil. It consists of two buckets, facing inwards and hinged together at the top like a clam or oyster shell.
It is dropped onto the soil in the open position and on raising it automatically closes around the soil to be excavated.

Claquage grouting
A form of grouting used mainly in fine-grained soil.
Tongues of high pressure grout penetrate in planes and zones of weakness within the soil, and in so doing compact the soil.
See also **Grouting**.

Classification of soil
See **Airfield soil classification**.

Clay
A natural deposit consisting of the finest products of rock weathering which forms a cohesive mass.
Clay has a smooth, greasy touch when squeezed between the finger and thumb. It dries slowly and remains plastic for a considerable time. Unlike silt it does not exhibit dilatancy. When dry a clay has considerable cohesive strength. A highly compressible clay is one which because of the fineness of its particle size and of its high moisture content can be considerably compressed under load. This type is also subject to large volume changes under the action of the weather.
Boulder clay may contain variable proportions of coarse material and large stones, but there is usually sufficient clay present to impart cohesion. Varved clays have thin alternative layers of sand or silt and clay. In particle-size analysis the clay grade consists of material of which the particle size is less than 0.002 mm. On the basis of strength the consistencies of clays may be classified as follows:

	Undrained shear strength kN/m²	*Consistency of clay*
Greater than	150	Very stiff
	100–150	Stiff
	75–100	Firm to stiff
	50– 75	Firm
	40– 50	Soft to firm
	15– 40	Soft
Less than	15	Very soft

Clay cutter
See **Hydraulic clay cutter**.

Clay fraction
The fraction of a soil composed of particles smaller in size than 0.002 mm diameter.

Clay grouting
Using a clay slurry to grout soil.
See also **Grouting**.

Clay-ironstone
See under **Claystone**.

Claypan
See **Hardpan**.

Clay puddle
'Plastic' clay used for water-proofing, such as the impervious cores in dams.

Clay sampler
See **Soil sampler**.

Clay spade
See **Graft**.

Claystone
A hard concretion of clay-ironstone or clay material cemented by calcite. Of flattened spheroidal form, hence the local name Turtle stone. Claystones may be over a metre in diameter and, when broken, they often show radiating cracks (septa) filled with calcite and are then called Septarian Nodules. They occur at certain horizons in clay formations such as the London clay, Oxford clay, etc, and in excavations appear arranged in lines apparently along a definite bed in the clay.

Cleading
See **Cladding**.

Clearance ratio
See **Inside clearance ratio**.

Clearing
Clearing a proposed construction site surface of all trees, brush, deadfall and surface boulders etc.
See also **Grubbing** and **Stripping**.

Cleat
A stop fixed to prevent movement of a strut or waling.

C-line
See under **Plasticity chart.**

Closed capillary fringe
See under **Capillary water.**

Closer
A sheet pile cut or made to close a cofferdam when a standard pile will not fill the gap.

Close sheeting
See **Close timbering.**

Close timbering
Planks placed vertically and touching each other against the ground. Used in 'running' ground.

Clubfoot roller
See **Sheepsfoot roller.**

Clunch
Hard chalk.

Coarse aggregate
See under **Aggregate.**

Coarse analysis
See **Sieve analysis test.**

Coarse grained soil
Soil containing not less than 90% passing a 37.5 mm BS test sieve.

Coarse gravel
Rock fragments ranging in size from 20 mm to 60 mm.

Coarse sand
Rock fragments ranging in size from 0.6 mm to 2 mm.

Coarse silt
A natural sediment of rock fragments ranging in size from 0.02 mm to 0.06 mm.

Coarse spoil
See under **Colliery spoil.**

Cobble gravel
See **Gravel.**

Cobbles
Rock fragments ranging in size from about 60 mm to 200 mm in diameter.
See also **Boulders.**

Cobi pile
A corrugated thin concrete cylindrical shell pile which is driven into the ground with an expanding mandrel which under compressed air grips the inside of the shell.

Code of Practice
A publication issued by the British Standards Institution describing what is considered to be good practice in a particular job. Such as CP 2001 on Site investigation.
See also **CP** and **BS.**

Coefficient of active earth pressure (K_a)
The ratio of the active earth pressure (p_a) normal to a plane surface to the corresponding pressure in a fluid of the same density.
In the original Rankine theory:—

$$K_a = \frac{1 - \sin \phi}{1 + \sin \phi}$$

where ϕ is the angle of shearing resistance of the retained soil.
See also **Coefficient of passive earth pressure.**

Coefficient of adhesion
See **Adhesion factor.**

Coefficient of compressibility (m_v)
The change in void ratio per unit increase of pressure per unit volume usually expressed in mm^2/N.
See also **Modulus of volume change.**

Coefficient of consolidation (C_v)
In the consolidation of soils a value expressed in mm^2/s, if the permeability is in mm per second.
It is equal to:

$$\frac{K}{\gamma_w m_v} = \frac{\text{coefficient of permeability}}{\text{density of water} \times \text{coefficient of compressibility}}$$

Coefficient of curvature
This refers to the shape of the soil distribution curve and equals $(D_{30})^2/D_{10} \times D_{60}$
For a well graded soil the coefficient of curvature must be between 1 and 3.

Coefficient of earth pressure at rest (K_o)
The ratio of the earth pressure normal to a plane surface to the corresponding pressure in a fluid of the same density, when there is no movement.
For sands K_o varies between 0.4 and 0.6 while for clays, at depth below the surface, it varies between 0.5 and 0.75.

Coefficient of imperviousness
An American term for the impermeability factor of a soil.

Coefficient of internal friction
The tangent of the angle ϕ_e, where ϕ_e is the angle of internal friction of a soil.

Coefficient of natural relative collapsibility
In the estimation of the value of settlement under the mass of the natural overburden is given by:

$$\frac{h_o - h'_o}{h_o}$$

where h_o is the height in millimetres of a soil sample at its natural water content loaded in a confined condition by a pressure equal to the pressure of its natural overburden,
and h'_o is the height, in millimetres, of the same sample after percolation of water through it.

Coefficient of passive earth pressure (K_p)
The ratio of the passive earth pressure (p_p) normal to a plane surface to the corresponding pressure in a fluid of the same density.
It is the reciprocal of (K_a) the coefficient of active earth pressure in the Rankine theory, that is:—

$$K_p = \frac{1}{K_a} = \frac{1 + \sin\phi}{1 - \sin\phi}$$

where ϕ is the angle of shearing resistance of the retaining soil.

Coefficient of permeability (k)
The average velocity of water through the total (voids and solids) area of soil under a hydraulic gradient of 1.
See also **Darcy's law**.

Coefficient of reduction
See **Reduction coefficient**.

Coefficient of secondary consolidation
This is given by the formula $C_t/ + e_o$
where C_t is the secondary compression index and is the slope of the straight line portion of the e log t curve obtained from a consolidation test,
and e_o is the voids ratio of the soil sample at the start of the test.

Coefficient of sorting (S_o)
This is a numerical value based on the particle size distribution curve of a sample of soil. It is obtained from the square root of the ratio of the quartiles; ie

$$S_o = \sqrt{\frac{D_{75}}{D_{25}}}$$

See also **Coefficient of uniformity**.

Coefficient of subgrade reaction
See **Modulus of subgrade reaction**.

Coefficient of thaw
See **Thaw coefficient**.

Coefficient of uniformity (C_u)
The ratio between the particle diameter which is larger than 60% by mass of the particles in a soil sample to that diameter, the effective size of which is larger than 10% by mass of the particles. It is expressed as D_{60}/D_{10}.
Uniform soils have a uniformity coefficient less than 3. Non-uniform soils have a relatively flat grading curve; uniform soils a steep one.
See also **Coefficient of sorting**.

Cofferdam
A structure, usually temporary, built for the purpose of excluding water or soil sufficiently to permit construction to proceed without excessive pumping and to support the surrounding ground. A land cofferdam is one constructed from a land surface; a water cofferdam is one constructed in open water.
The various types of cofferdam that can be constructed are as follows:— **Cantilever, Cellular, Circular Arc, Circular cell, Double skin, Double wall, Earth fill, Gravity, Ground stabilised, Ring type, Rock fill, Sheet pile, Single skin, Sheet steel** and **Timber crib**.
See also **Box dam, Caisson** and **Closer**.

Cohesion
Mutual attraction that exists between the fine particles of some soils and thus tends to hold them together in a solid mass without the application of external forces.

Cohesion assessment
See under **Rapid assessment of plasticity**.

Cohesion number
A non-dimensional number equal to

$$c/\gamma z$$

where c is the soil cohesion,
γ is soil density,
and z is depth.
See also **Stability number**.

Cohesive soils
These consist of the finer and altered products of rock weathering. In their natural state they possess cohesion and plasticity, such as in clay.

Cohron sheargraph
A hand held device designed to obtain information on the *in situ* strength properties of soil as well as on the shearing resistance between soil and metal and soil and rubber.
In operation, the small diameter shearhead is completely inserted into the soil, normal stress is applied to the shear surface through axial deflection of a spring, and shearing stress is applied by twisting a recording drum until the soil fails. After soil shear failure occurs, normal load is gradually reduced. Since the soil will sustain only a given amount of shearing stress for a particular normal load, the recording pen will trace a curve of shear stress versus normal stress as the latter is reduced to zero.
Soil to metal or soil to rubber shearing resistance is determined by inserting a smooth metal or rubber faced shearhead in the device.

Co-ions
Ions present in the surface layer, surrounding a particle, with the same polarity as is possessed by the colloid.

Collapsible soil
A description given to soil which rapidly decreases its voids ratio on the introduction of water. Such soils occur in many parts of the world where there is a long dry season and where the water table is at great depth.

Colliery spoil
Discarded material from a colliery, which if suitable can be used as a 'fill' material.
Coarse spoil consists of normal excavated material and reflects the various rock types encountered. Fine spoil consists of either slurry or tailings from the colliery washery.

Colloidal micelle
A term sometimes used to describe a colloidal particle plus its double layer solution.

Colloidal solution
A suspension in which the particles do not settle because they respond to molecular movement in the fluid.

Colloids
Particles smaller than 0.002 mm (the European definition of the largest of the clay particles) and larger than 0.000001 mm, ten times the diameter of an atom. Particles smaller than a diameter of 0.002 mm do not settle in water and those between 0.002 and 0.0002 mm settle only very slowly.

Colluvial soil
Any soil which has been deposited mainly through the action of landslides and slopewash.

Colorimetric pH equipment
See under **Barium sulphate test**.

Columnar structure
See **Fissured**.

Combined footing
A footing supporting two or more columns.
See also **Trapezoidal footing.**

Compact
Compact non-cohesive material. In quantitative terms, a material may be regarded as compact if its relative compaction exceeds 90%, ie if its dry density exceeds 90% of the maximum dry density obtained in the standard compaction test.

Compacted fill
See **Controlled fill.**

Compacted fill density
See under **Compaction.**

Compaction
The density at which soil can be placed in an earth dam, embankment or road sub grade depends on its moisture content and on the amount of energy which is used in compaction.
See also **Hydrocompaction.**

Compaction curve
See under **Proctor compaction test.**

Compaction machine
See **Automatic compaction machine.**

Compaction mould
See **Proctor mould.**

Compaction permeameter
A cell which is basically a Proctor mould which is clamped between a base plate and top cap so that a falling head permeameter test can be carried out on a compacted sample of soil.

Compaction rammer
A 2½ kg rammer, having a 300 mm drop, used to compact soil in a Proctor mould before carrying out a standard compaction test.
See also **AASHO rammer**.

Compaction test
See **Proctor compaction test.**

Compensated foundation
Constructing the foundation slab of a structure at such a depth below existing ground level that the mass of the structure and foundation is balanced by the mass of the soil excavated.

Composite pile
A pile having one material for the initial driving into the ground and an added upper portion of a different material.
See also **Fuentes pile, MacArthur pile, Palm pile, Rotinoff pile** and **Thermopile.**

Compozersystem
A system of ground improvement developed in Japan in 1957.
The system consists of sinking a thick-walled steel tube with a vibrator on the top into the subsoil to the required depth, if necessary with the aid of water and air jets. A sand plug at the lower end of the tube prevents penetration of the soil into the tube. At the required depth a certain quantity of sand is placed inside the tube after which the tube is withdrawn one metre. During the withdrawal sand flows out of the tube aided by air pressure that has been introduced on top of the sand. After withdrawal over this length the tube is sunk again partially, pushing the freshly deposited sand into the surrounding soil with the assistance of vibrations.
The ratio between the steps of extraction and redriving governs the cross-section of the sandpile. By repeating the steps gradually a high density pile is thus built up.
See also **Vibroflot**.

Compressed air
Air that is compressed to a greater pressure than one atmosphere. It is pumped into underground workings to stop the ingress of water and mud into the working chamber. Men and materials must then pass into and out of the chamber through an air lock.

Compressed-air caisson
A caisson with a working chamber in which the air is maintained above atmospheric pressure to prevent the entry of water and ground into the excavation.

Compressed pile
A steel tube is placed vertically where the pile is to be driven. Dry concrete is placed in the bottom of the tube and a drop hammer delivers blows on the concrete until it forms a plug which penetrates the soil and takes the tube down with it. When the required depth is reached, the tube is slowly raised and the concrete forced downwards and outwards by the drop hammer until a bulb of concrete is formed. A reinforcing cage is introduced into the tube and successive quantities of concrete and ramming inside the cage are carried out until ground level is reached.

Compressibility
Volume change in a soil mass due to changes in stress caused by both natural and artificial means.
See also **Coefficient of compressibility**.

Compressibility modulus
See **Modulus of compressibility**.

Compression index (C_c)
The tangent of the slope angle of the straight part of the field consolidation line for clay soil.

Compression

It is obtained from the equation:

$$e = e_o - C_c \log_{10} \frac{p_o + \Delta p}{p_o}$$

where e_o is the initial voids ratio of the soil which corresponds to the initial pressure p_o,
and Δ_p is a small change in pressure.

Compression modulus of rubber
See under **Rubber extension modulus.**

Compression pile
A pile which is designed to resist an axial force such as would cause it to penetrate further into the ground.
See also **Bearing pile.**

Compression test
See **Consolidation, Triaxial compression** and **Unconfined compression test.**

Compression test machine
A basic load-applying machine which can be adapted to carry out tests such as the unconfined compression test, triaxial compression test, CBR test etc.

Concrete pile
A pile made with concrete. The main categories are *in situ* cased piles, *in situ* uncased piles and precast piles.

Concretions
Accumulations of carbonates or iron compounds found in clays or shales. They take the form of roughly spherical nodules and vary in size from a few mms to a metre.

Conducting paper
Specially prepared electrical conducting paper used to work out seepage problems.
See also **Electrical seepage analogue.**

Conductor
The first casing of a cased borehole. It is often secured in position by surrounding it with concrete.

Cone foundation
A foundation, to support a column, in which the underlying soil is first shaped into a cone and then covered by a uniform thickness of reinforced concrete.

Cone of depression
When water is pumped from a well point the water table is lowered adjacent to the point to form, after a suitable time has elapsed, a cone of depression.
See also **Pumping in test.**

Cone of radiation
See under **Ground probing radar.**

Cone penetration test
The testing of soils by pressing a standard cone into the soil under a known load and measuring the penetration.
See also **Cone penetrometer apparatus.**

Cone penetrometer apparatus
A new apparatus to measure the liquid limit of a soil.
See also **Liquid limit device.**

Confined compression test
See **Compression test** and **Triaxial compression test.**

Confined water
Water in a permeable stratum of soil that is confined between impermeable strata.
See also **Semi-confined water.**

Confined well
See under **Well.**

Confining pressure
The cell water pressure in a triaxial compression test.

Conglomerate
A rock consisting of rounded pebbles and boulders of other rocks in a cemented matrix, forming a consolidated gravel.
See also under **Breccia.**

Connate water
Water trapped in sediments at the time of deposition.

Consistency
The degree of resistance of a fine-particled soil to flow or to deformation in general.

Consistency Index (CI)
A figure for comparing the stiffness of clays in their natural state. It is calculated as follows:

$$\frac{(\text{liquid limit}) - (\text{water content} \times 100\%)}{(\text{liquid limit} - \text{plastic limit})}$$

It may rise above 100% but such values indicate a dry, friable and crumbly clay.
See also **Liquidity Index.**

Consistency limits
The liquid limit, plastic limit and shrinkage limit. These are all water contents of a clay, each in a certain condition defined by BS 1377. They are the standard way of describing clays.
See also **Atterburg limits.**

Consistent condition
A term used in boring or tunnelling to describe conditions that remain constant throughout the entire length of the bore.
See also **Erratic condition.**

Consolidated quick test
A consolidated undrained test.

Consolidated undrained test
A shear test of a soil sample carried out in a triaxial cell. Drainage of the sample is allowed during the slow application of the all-round stress so that the sample is fully consolidated under this pressure. During the application of the deviator stress no drainage of the sample is allowed. The test can be carried out on undisturbed samples of clay, peat and silt, on remoulded samples of clay and silt, and on redeposited samples of cohesionless soil.
See also **Undrained test**.

Consolidation
The gradual compression of a cohesive soil, due to a mass acting on it, which occurs as water is driven out of the voids in the soil. Consolidation occurs only with clays, or other soils, of low permeability. It is not the same as compaction which is an artificial mechanical process.
See also **One-dimensional, Secondary consolidation** and **Three-dimensional consolidation**.

Consolidation cell
A circular cell containing removable upper and lower porous plates between which the soil specimen under test is placed. Above the upper porous plate is a circular loading plate which can slide vertically inside the cell. The cell is rigidly fixed to a base plate. Integral with the cell is a water reservoir to ensure that the soil specimen under test is fully saturated at all times.
See also **Rowe consolidation cell**.

Consolidation coefficient
See **Coefficient of consolidation**.

Consolidation settlement
The settlement of loaded soil which takes place over a period of years.
It is due to the dissipation of excess pore pressure as water, or air and water, are expelled from the voids in the soil, with a consequent decrease in volume. This consolidation will be slow in soils of low permeability.
In some cases further settlement in the nature of a long-term creep may occur. This consolidation can sometimes be accelerated by vertical sand drains. The leaning tower of Pisa is an example of unequal consolidation settlement.
See also **Immediate settlement**.

Consolidation test
A confined compression test carried out in an oedometer to investigate the relationship between the applied vertical pressure and time of settlement of a sample of soil.

Consolidometer
See **Oedometer**.

Constant head permeameter
See under **Permeameter**.

Constant head permeability unit
A prefabricated unit consisting of a constant head mercury pressure unit which

pressurises the de-aired water in one tube of the piezometer through a column of paraffin, the other tube from the piezometer remaining connected to the manometer recording panel. The paraffin-water interface passes through a specific sized burette in order to measure the amount of water flowing to the piezometer tip.

Constant pressure device
See **Pressure system.**

Constant rate of penetration test (CRP)
A test in which a pile is made to penetrate the soil from its position as installed at a constant speed while the force applied at the top of the pile to maintain the rate of penetration is continuously measured.
See also **Constant rate of uplift test.**

Constant rate of uplift test (CRU)
A test in which a pile is extracted from its position as installed in the soil at a constant speed while the force applied at the top of the pile to maintain the rate of uplift is continuously measured.
See also **Constant rate of penetration test.**

Constant volume test
To ensure complete saturation of the soil sample in a standard consolidated, undrained test on a fully saturated soil sample it is necessary that the volume of the sample is held constant during shear by varying the cell pressure throughout this stage of the test so as to maintain a constant pore pressure under conditions of no drainage.

Contact angle
The angle between the surface of water in a capillary tube and the wall of the tube.

Contact moisture
Capillary moisture in a soil above the water table where there is more air in the voids than water, and the water is discontinuous.

Contact pressure
The pressure per unit area exerted by a foundation on the soil immediately below it.
See also **Capillary pressure.**

Contact spring
See **Stratum spring.**

Continental shelf
The shallow, submerged platform, bordering and marking the structural edge of a continent. The term 'shelf' refers to the sea bed to a maximum depth of 200 metres.

Continuous back drain
See **Back drain.**

Continuous flight auger
See **Helical auger.**

Continuous footing
A footing below a wall or a line of closely spaced columns.

Continuous helical auger
See **Helical auger**.

Continuous soil sample
See under **Liner**.

Contour line
A line drawn on a map between all points at the same level.

Controlled fill
Suitable 'fill' soil or inert material placed on natural ground in thin layers under controlled compaction after any existing weak and compressible soil has been removed.
See also **Compaction** and **Pass**.

Convergence measuring device
See **Tape extensometer**.

Converse–Labarre formula
A formula to give the efficiency of a group of piles with respect to their overall bearing capacity:

$$\text{Efficiency} = 1 - \frac{\phi\,(n-1)m + (m-1)n}{90\ mn}$$

where m = number of rows,
n = number of piles in a row,
ϕ = D/S in which ϕ is numerically equal to the angle whose tangent is D/S, degree,
S = spacing centre to centre of piles in mm,
and D = Diameter of pile in mm.

Co-ordination number
The max. number of anions which can surround a cation to form a stable group.

Core
An undisturbed cylindrical soil or rock sample. Cores can vary in diameter from 30 mm to 150 mm and be up to 1.5 m in length.

Core box
A large wooden box divided into narrow strips by partitions. It is used to place the cores from a borehole in the order in which they were obtained.

Core catcher
A series of sprung truncated flaps at the bottom of a sampling tube which enables a sample of cohesionless soil to be obtained.
See also **Flapper valve**.

Core cutter
See **Core drill**.

Core drill
An annulus drilling bit that leaves a central core of undisturbed soil or rock.

Core lifter
A lifting tube acting inside a drilling tube which enables a soil or rock core sample to be brought to ground level through the inside of the drilling tube.

Core wall
See **Cut off wall.**

Coring
See **Predrilling.**

Cotton soil
See **Black cotton soil.**

Coulomb, Charles *(1736–1806)*
French army engineer. In 1773 he published a paper on the earth pressure against retaining walls, which today is still the basis for the design of retaining walls.

Coulomb equation
See under **Angle of shearing resistance.**

Coulomb line
See **Envelope of failure.**

Coulomb wedge theory
See **Wedge theory.**

Counter drain
A drain running along the toe of a slope.

Counterfort drain
See **Buttress drain.**

Counterfort wall
A retaining wall consisting of a thin face slab supported at intervals on the inner side by vertical counterforts that meet the face slab at right angles, the whole being supported on a base slab.

Counter ion
An ion of opposite charge, in the solution surrounding a particle, which counterbalances the charge on a particle.

CP
Symbol for a Code of Practice such as CP 2001 on Site Investigation and CP 2003 on Earthworks.
See also **BS.**

Crack
See **Joint.**

Creep
To move slowly. Soil creep is sometimes observed on slopes, where it is mainly caused by gravitational pull.

Creep settlement
The settlement of soil which occurs following dissipation of the excess pore water pressure.

Creep strength
The shear strength of undisturbed clay which, if not exceeded, exhibits creep upon the applicaiton of shearing stress and thereafter experiences no progressive deformation. If the creep strength is exceeded, the clay deforms continuously under constant shear stress.

Crest line
See under **Watershed**.

Crib (or **grillage**)
A layer of timber, old rails or steel joists in either one or two directions placed beneath a shore, post, or column, for the purpose of spreading the load.

Crib wall
A retaining wall constructed of rectangular interlocking precast concrete or timber members to form a cellular structure, laid on top of each other, and filled with soil or broken rock.

Cribwork
The making of crib-type structures.

Critical circle
The slip circle of failure of a soil slope which has the minimum factor of safety.

Critical density
The density of sand at which no change in volume is brought about upon the application of shear.

Critical height of slope
The vertical height of a slope for a factor of safety of unity.

Critical hydraulic gradient
The hydraulic gradient at which the soil becomes unstable, ie when the intergranular pressure becomes zero.

Critical porosity
Porosity of sand at its critical density.

Critical pressure
The pressure at which the cohesive bond between soil particles begins to fail.

Critical slip surface
See **Critical circle**.

Critical state design
In soil mechanics it is analogous to plastic design in steel frame structures.

Critical voids ratio
The voids ratio of a sand at its critical density.

Cross bedding
See **Current bedding**.

Cross poling
In excavations, short lengths of poling boards placed horizontally across a gap between runners or sheeting and tucked in behind them and used where runners or sheeting cannot be driven continuously and vertically.

Crowd shovel
See **Face shovel**.

Crowsfoot
A square shanked steel rod bent at its bottom end into a horizontal U shape. It is fixed to the bottom end of a string of boring rods to recover 'lost' boring tools.

CRP
Constant rate of pile penetration test.

CRU
Constant rate of pile uplift test.

Cruciform vane
A four-bladed vane in the shape of a crucifix.

Cryogenics.
The study of soil, or other material, at very low temperatures.
See also **Frozen soil**.

Cryogenic sampler
A core sampler which uses liquified nitrogen to freeze the *in situ* soil sample to be extracted.
This type of sampler is used to extract soil which would otherwise be difficult to sample without serious disturbance.

Culvert
A covered channel, or a large diameter pipe, for carrying a water course below ground level, mainly under a railway or road.

Cup auger
An auger designed by John Ranttila to remove soil from a sample tube in small increments of controlled height and volume.
In operation, the auger is lowered to the top of the sample and a stop collar is clamped at a graduation mark a short distance above the cap. The auger is rotated until the stop collar is in contact with the cap, whereupon it is withdrawn, emptied, and again inserted for final trimming, when the entire operation is repeated.

Current-bedding (also known as **cross-bedding**, and as **false-bedding**)
Bedding oblique to the planes of general stratification, caused by swift local currents, frequently seen in sandstones, river sands and gravels, delta deposits and sand dunes.

Curvature coefficient
See **Coefficient of curvature**.

Curvature pressure
See under **Capillary pressure**.

Cushion
A layer of gravel-sand, equal in depth to the width of foundation, placed beneath a foundation to cushion low-bearing capacity soil below.
See also **Dolly**.

Cut and cover
An alternative to tunnelling, where a trench is excavated to formation level and the tunnel built in the open. The tunnel is then covered over.

Cut and fill
A description given to the construction of a canal, railway or road where they are partly embanked and partly below ground in cut.

Cut-off depth
The depth reached by sheet piling or cofferdam walls below excavation level.

Cut off trench
See under **Cut off wall**.

Cut off wall
An impervious wall built to prevent the movement of water through the soil; such as under a dam, along a river bank, surrounding a gravel pit or around the periphery of an excavated construction site. Cut off walls are particularly useful in situations where permeable strata overlie an impermeable stratum into which the wall may be founded.

Cutting
An excavation for carrying a canal, railway or road below ground level in the open.

Cutting-out piece
A short piece of 'timber' which may be cut out to facilitate the striking of 'timbering'.

Cutting ring
A 25 mm deep by 75 mm interval diameter stainless steel or brass open-ended ring having a cutting edge along the bottom of the ring to enable an undisturbed sample of cohesive soil to be cut from a larger sample of soil, or the end of a U4 tube of undisturbed soil, to be tested in an oedometer.

Cutting shoe
An externally chamfered cutting edge which is fixed to the bottom of a soil sampler or U4 tube. It has the same internal diameter as the sampler but has a larger external diameter so that an air gap is formed around the sampler. This gap stops a vacuum being formed on withdrawal of the soil sample.

Cutting square
A 20 mm deep by 60 mm sided square open-ended stainless steel box having a cutting edge along the bottom of each side to enable an undisturbed sample of cohesive soil to be cut from a larger sample of soil, or the end of a U4 tube of undisturbed soil, to be tested in a 60 mm square shear box apparatus.

Cycle time
A term used to describe the time taken by a scraper to complete a complete cycle of loading, hauling and returning to reload.
See also **Travel time**.

Cyclical load test
A test to estimate the distribution of the axial load along the length of a pile.

Cyclotherm
A multiple repetition of beds of different lithology which are recognisably similar in internal sequence, showing usually minor variations both in thickness and precise sequence of components.
For example, in the United Kingdom coal measures the simple cyclothermic unit usually consists of coal, seatearth, sandstone, non-marine shale or mudstone, marine band and then coal etc, again.

Cylinder
An alternative name for an open caisson or monolith of cylindrical form.

Cylinder penetration test
A comparative test of the quality of stabilised soil.

Cylindrical piezometer tip
See under **Piezometer tips.**

Cylindrical slide
See **Rotational slide.**

D

D_{10}
Symbol for the effective size of a soil sample.

Dam
See **Earth dam.**

Damage index
See **Index of damage.**

Damp proofing
Providing a waterproof barrier around, in or under a building structure to prevent the ingress of ground water.
See also **Electro-osmosis.**

Darcy's law
When the hydraulic gradient is equal to or less than unity, the discharge velocity, v, under the action of an hydraulic gradient, i, is given by the equation

$$v = ki,$$

where k is the coefficient of permeability.

Dead load
The permanent mass of a structure and its fixings.

Deadman
An anchor plate.
See also **Anchor block.**

Dead shore
A vertical shore used for temporarily supporting the upper parts of a wall, the lower parts of which are required to be removed in the process of underpinning.

De-aired water
Water that has been boiled under a vacuum.

De-airing unit
A special unit for making and storing de-aired water, such as required in pore water pressure measurements.

Decompaction
Decrease in soil density, usually brought about by over-vibration.

Deep compaction
See under **Vibroflot**.

Deep foundation
A foundation founded at a depth usually greater than 3 metres.
See also **Shallow footing**.

Deep pit
See under **Pit**.

Deep trench
See under **Trench**.

Deflected pile
A pile which has deflected away from its intended direction during driving. This is usually caused by the pile hitting an obstruction.

Deflection inclinometer
See **Inclinometer**.

Deflectometer
See **Inclinometer**.

Deflocculants
Agents to promote the dispersion of flocks, or solid particles which are coalesced into larger groups.
See also **Grouting**.

DEGEBO
Deutsche Gesellschaft für Bodenmechanik. German Society for Soil Mechanics.

Degree days
See under **Freezing index**.

Degree of compaction
The tightness of packing of a soil sample, estimated by the formula:—

$$\frac{(\text{voids ratio in loosest state}) - (\text{voids ratio of sample})}{(\text{voids ratio in loosest state}) - (\text{voids ratio in densest state})}$$

Degree of consolidation (U_v)
The settlement after time t divided by the final settlement.

Degree of disturbance
See **Area ratio**.

Degree of saturation (S_r)
The percentage of the volume of water in the voids of a soil to the total volume of voids between the soil particles.
It gives a measure of air (or gas) in the voids since the air content is 100% minus the degree of saturation.

Degree of shrinkage
This is given by the formula

$$\left(\frac{V_i - V_f}{V_i}\right) \times 100\%$$

where V_i is the initial volume of the soil sample before drying,
and V_f is the final volume of the soil sample after drying.

Dehottay process
A refinement of the ground freezing process for shaft sinking or foundations. Instead of brine being circulated in the pipes installed in the ground, liquid carbon dioxide is pumped in. An advantage of carbon dioxide is that when it passes out through a leak in the pipe into the freezing ground it has no effect on its freezability. If brine escapes, the freezing point of the ground is so much lowered that it may be impossible to freeze it again.

Dehydration test
This test consists in recording the percentage loss in a mass of clay upon heating to higher and higher temperatures and plotting it against the temperature. On any particular temperature, heating is continued until no loss in mass occurs. The test is continued until a temperature is reached at which there is no further loss in mass on further increase of temperature.
The position of the flexural point in the temperature-percentage loss in mass curve gives an indication of the type of mineral present.

Delft sampler
A hollow piston-type soil sampler which can take a continuous borehole sample of 29 or 88 mm diameter from ground level to a depth of 20 m. The sample core of soil is contained within a self-vulcanising sleeve as the sampler is pushed further into the ground.
The Swedish foil sampler performs the above function in retaining a continuous core, but within an aluminium foil.

Delta deposit
See **Alluvial fan.**

Demountable reference studs
See under **Reference studs.**

Dennison sampler
A double tube core barrel soil sampler having a disposable liner 150 mm diameter and 600 mm long for handling and transporting the soil cores. It is suitable for sampling fine-particled soils and will recover reasonably undisturbed samples if the soil is slightly cohesive and the boring done carefully. See also **Denver sampler.**

Densification
See **Compaction.**

Density (γ)
The density of a soil is its mass per unit volume.
See also **Bulk density, Dry density, Partially saturated density, Saturated density** and **Wet density**.

Density bottle
A small glass bottle of specific volume which has a ground surface on the inside of its neck and a close fitting ground surface stopper with a central capillary vent.
See also **Pycnometer**.

Density chisel
A hardened tip cold chisel for removing highly compacted material from a 'density' hole.

Density index
See **Relative density**.

Density probe
A pressure-tight stainless steel cylinder containing a radioactive source which is pushed into the soil. When used with a surface sensing device the *in situ* density of the soil can be found.
See also **Moisture probe**.

Denver sampler
A double tube core barrel soil sampler having a disposable liner 143 mm in diameter and 711 mm long for handling and transporting the soil cores.
The sampler consists of a rotating outer barrel with cutting teeth on the bottom, a non-rotating inner barrel with a smooth cutting shoe, a spring operated core catcher and a liner to receive the sample.
In operation, the inner barrel remains stationary and slides over the cylindrical shaped sample which is cut by the rotating outer barrel.
See also **Dennison sampler**.

Deposit
A sedimentary layer of soil.

Depression cone
See **Cone of depression**.

Depth factor
A factor by which the vertical height of a slope has to be multiplied to give the depth of the lowest point of the slip circle failure surface below the top of the slope.

Depth measurement transducer
See **Displacement transducer**.

Depth of thaw
See **Thaw depth**.

Derrick
The structure used to support the auger and other equipment which has to be raised or lowered during the drilling of a borehole.

Detrimental settlement
Settlement or cracking of a structure due to the stress-deformation of the underlying soil.

Deviatoric plane
The plane through the origin perpendicular to the space diagonal which has the principal stress equation

$$\sigma_1 + \sigma_2 + \sigma_3 = 0$$

Deviator stress
The difference between the soil sample failure stress and cell pressure in a triaxial compression test, ie the diameter of the Mohr circle of stress at failure.

Dewatering
See **Ground water lowering**.

Diagenesis
The modification of clay after it has been mixed or leached with water containing electrolytes.

Dial gauge
An instrument which shows, by a needle indication on a circular graduated dial, very small displacements of its plunger. The plunger is accurately geared to the needle. In soil mechanics it is often used in conjunction with a proving ring.

Diamond vane
A crucifix vane of diamond shape so that the plane of failure measured is at an angle to the horizontal. This shape is considered to be more suitable than the normal square or rectangular shaped crucifix vane for measuring the shear stress on a slip circle failure of a slope.
See also **Vane test**.

Diaphragm pump
See **Sludge pump**.

Diaphragm wall
The technique of constructing *in situ* (ie, in the place of its use) a separating wall in the ground. The wall is generally of mass or reinforced concrete.
The technique involves the excavation of a deep narrow trench using a bentonite/water suspension to replace the excavated soil. This suspension is later replaced by concrete.
An essential feature of construction is the digging of a pilot trench one metre deep along the line of the proposed diaphragm wall. This trench is faced with concrete and should be about 25 mm wider than the special excavating grab. It sets out the line and level of the work and guides the excavation. It also acts as a reservoir for the bentonite suspension.
The excavator is set up over the guide trench and the soil excavated to the full required depth over the length of one grab bite. As excavation proceeds the bentonite suspension is pumped into the hole to replace the removed soil. For practical purposes a continuous wall is constructed in sections varying between 1.5 m and 6 m in length. Concrete is placed by tremie to the bottom of the trench in a continuous pour, displacing the bentonite suspension, which is pumped away to waste or to a storage for re-use.

Diaphragm

Wall sections are joined by use of a steel tube which is inserted as an endshutter or stopend in each bay. As the concrete sets the tube is withdrawn, leaving a vertical construction joint against which the adjacent wall panel is cast.

Diaphragm wall cofferdam
See under **Cofferdam**.

Diatomaceous earth
A fine, light grey, soft sedimentary deposit of the siliceous remains or skeletons of minute unicellular marine organisms. Such a soil is very porous, and of fine structure.

Differential settlement
Where the foundations of different parts of the same structure settle at different rates.

Digital indicating system
A system which monitors performance of soil tests by using normal digital display calibrated indicators.
See also **Signal conditioning module** and **'Dot' recording system**.

Dilatancy
Resistance to shaking.
See also under **Shaking test** and **Silts**.

Dimension stone footing
An old type of footing consisting of stone cut and dressed to size by a mason.

DIN
Deutsche Industrie Normen (German industrial norms).

Dipmeter
An instrument to record the depth below ground level of the surface of the water in a borehole or piezometric tube. It consists of an electric cable mounted on a winding reel which contains an amplifier and oscillator which actuates to give a high pitched note when the water level is reached.

Direct shear test
See **Shear box**.

Direct transmission density
See under **Nuclear densometer**.

Dirty money
Additional pay to a construction worker for working in difficult or dirty conditions.

Disc auger
This consists of a single helical disc, up to a diameter of one metre, fixed to the bottom end of a drilling shaft.
The lower end of the single helical has a cutting edge, while the upper end has a hinged flap to contain the soil once it has been cut away.

Discharge velocity
The quantity of liquid that percolates in unit time across a unit area of a section of soil at right angles to the direction of flow.
See also **Darcy's law**.

Discharge well
See under **Well**.

Disc load cell
A mechanical load cell which has an elastic element consisting of a cup spring. The element, fixed between an abutment plate and a top yoke plate, deflects when loaded, and a dial gauge measures the distance between the abutment and yoke plates. The whole is contained in a watertight steel peripheral ring.

Disc piezometer tip
See under **Piezometer tips**.

Dispersed structure
A description of the arrangement of clay particles where the edges, corners and faces of the clay platelets have like electrical charges such that the particles repel each other and assume nearly parallel positions.

Dispersing agent
A deflocculating agent, such as sodium oxalate, used in wet analysis to prevent soil particles joining together and settling more quickly than they normally would.

Dispersion test
A test for distinguishing between silt and clay and for estimating the relative amounts of clay, silt and sand in a soil sample.
A small quantity of the soil is dispersed with water in a glass cylinder and then allowed to settle. The coarser sand particles settle first and the finest particles remain in suspension the longest.

Displaced pile
See **Deflected pile**.

Displacement transducer
A transducer which converts displacement to a proportional electrical signal. For instance, the transducer can replace the dial gauge measuring the displacement of a proving ring, as used in soil testing.

Displacement vessel
See **Water displacement vessel**.

Dissipation test
A test to study the build up of pore pressure in a soil sample due to a stress change under undrained conditions. Also the subsequent rate of dissipation of the pore pressure when drainage is permitted from one end of the sample under test. The test is conveniently carried out in a triaxial compression cell.

Dissolved salts
See under **Intergranular cement**.

Distribution curve
See under **Sieve analysis test**.

Distributor
See **Cap**.

District Surveyor
See under **Building inspector**.

Disturbance area
See under **Area ratio**.

Disturbed sample
See under **Sample**.

Ditcher
The American name for a trench cutting machine.

Dog
A fastening of iron used for spiking large excavation timbers together and having both ends bent down and pointed.

Dog bar
See under **Bar dog**.

Dogleg
A sharp bend in a borehole, either deliberate or accidental.

Dolly
A block of hardwood, or other suitable material, placed on the top of a concrete or timber pile to cushion the blows from a pile driving hammer.
See also **Long dolly**.

Dome foundation
A foundation, to support a column, in which the underlying soil is first shaped into a dome and then covered by a uniform thickness of reinforced concrete.

Dörr's formula
The safe bearing capacity for vertical-sided piles as derived by Dörr.

'Dot' recording system
A unit, used in conjunction with an analogue or digital indicating system, which can directly produce a graph showing any required linear displacement, load or pressure change, plotted against time in a laboratory soil test.

Double layer
The electrical charge on the surface of a colloidal particle together with the charge due to the balancing counter ion.

Double oedometer test
See **Swelling test**.

Double shaft pier
Twin piers separated by a joining structure.

Double skin cofferdam
See under **Cofferdam** and **Double wall cofferdam**.

Double tube core barrel
See **Core lifter**.

Double-wall cofferdam
A cofferdam enclosed by a wall consisting of two parallel lines of sheeting tied together and with filling between them and which is usually self-supporting against external pressure.

Downdrag
See **Negative skin friction.**

Downdraw
Lowering the water level in a reservoir.

Dowsing
Searching for water, and ore deposits, by the feeling of a branch or a pendulum held in the hand.

Dozer
See **Bulldozer.**

DPC
Abreviation for damp proof course.

Drag
(1) The American name for Negative skin friction.
(2) A machine fitted with two or more oblique blades for scraping off and reshaping irregularities in the surfaces of earth roads.

Dragline
This consists of a long boom mounted on an excavator and a specially designed bucket. It operates by swinging its bucket out onto the material to be excavated, then hauling it back towards the base machine, excavating and filling as it comes in. The machine is designed to stand above its work and to move backwards as excavation proceeds.
See also **Walking dragline.**

Drag shovel
See **Back acter.**

Drain
See **Perforated drain.**

Drainage
The process of transferring liquid from one location to another location in or out of the soil.

Drainage blanket
A layer of pervious gravel or sand laid over existing relatively impervious soil, in which vertical sand drains have already been installed and on which an embankment can be constructed. The combined sand drains and blanket allow water to be squeezed out of the subsoil as the mass of the embankment is increased and this increases the rate of consolidation of the subsoil.

Drainage path
The maximum distance that a pore water particle under pressure, in a particular soil stratum, has to travel.
In a soil stratum bounded with pervious soil strata above and below it the maximum distance of travel is one half the thickness of the stratum. This is known as two-way drainage.
In a stratum bounded with one pervious layer and one impervious layer the maximum drain path is equal to the stratum thickness and this is known as one-way drainage.

Drained angle of internal friction
See **Angle of internal friction**.

Drained test
A shear test of a soil sample, usually carried out in a triaxial cell.
Drainage of the soil sample is allowed throughout the test, so that full consolidation occurs under the application of an all-round stress and no excess pore pressure is set up during the application of the deviator stress.
Drained tests can be done on compacted, redeposited, remoulded or undisturbed samples of soil, which can be either fully or partially saturated.

Drain hole
See **Weep hole**.

Drain pile
See **Vertical sand drain**.

Drain tile
The American name for an agricultural drain.

Dredge bulkhead
A bulkhead which, after being anchored, has soil excavated or dredged from the front of its base.
See also **Fill bulkhead**.

Dredge level
The level to which the ground on the water side of a bulkhead, wharf or quay has been dredged.

Drift
(1) Amongst British geologists this term is, today, used to cover all superficial deposits of the earth's crust, such as boulder clay, sands and gravels, brick-earth, head, alluvium and the like.
(2) The horizontal distance by which a non-vertical borehole or driven pile is displaced at depth from the point vertically below the top.

Drilled caisson
See **Drilled pier**.

Drilled pier
A pier formed by pouring concrete into a large diameter bored hole. The pier can be straight shafted or under-reamed.

Drilling fluid
See **Drilling mud**.

Drilling machine
See **Rig**.

Drilling mud
A slurry of clay and water to which bentonite is usually added. It is used during the drilling of a borehole to prevent the collapse of the hole as well as to seal off permeable strata.

Drilling rig
See **Rig**.

Drilling rod
See **Boring rod**.

Drive-in piezometer tip
See under **Piezometer tips**.

Drive-in standpipe piezometer
See under **Standpipe piezometer**.

Driven cast-in-place pile
A pile composed of a driven permanent steel or concrete casing which on reaching the correct depth is filled with concrete.

Driven pile
A pile driven into the ground by the blows of a hammer or by a vibrator.

Driving band
A steel band fitted round the top of a timber pile before driving into the ground.

Driving cap
A temporary steel cap placed on top of a steel pile to distribute the blow over the cross-section and to minimise damage to the head during driving.

Driving resistance
See **Set**.

Drop hammer
The simplest form of pile driving hammer. It consists of a cast-iron block provided with a lifting eye and side projecting lugs which slide in the pile frame leaders. The hammer is lifted by a winch for each blow and falls back under its own mass onto the pile.

Drop shaft
See **Open caisson**.

Dry density (γ_d)
The mass of solids contained in a unit volume of the soil.
See also **Density**.

Dry strength
Crushing characteristics.

Dry well
The American name for a soakaway.

Dumper
A rubber-tyred vehicle with two large wheels in front and two smaller wheels behind, with a hopper positioned over the front wheels. This hopper usually dumps forward but on some models can also dump sideways.

Dumpling
The ground temporarily left in the middle of an excavation which may serve as an abutment for the 'timbering' to the surrounding trenches.

Dune sand
See **Sand dune**.

Durichlor
A special type of sacrificial anode for use in silt.

Dutch auger
An auger specifically designed to bore through fine-particled soils.

Dutch cone
A penetrometer developed in Holland to determine the depth of a good bearing stratum below loose overlying soil.
The penetrometer is cone shaped with a maximum area of 1000 mm^2. The cone is attached to a rod which can slide in an outer sleeve.
The force needed to drive the cone and sleeve into the soil may be measured independently, so that the end bearing resistance and side friction may be separately determined.

Dyeline
A print with dark lines on a near white background.

Dynamic consolidation
Compacting soil by dropping in free fall a mass of up to 20 tonnes, from a height varying between 6 m and 25 m, on to the ground to be compacted.

Dynamic penetration test
A test carried out by driving a penetrometer into the soil. It is usually used for measuring the resistance of hard soils.
See also **Penetration tests** and **Static penetration test**.

Dyne
A dyne is that force acting on a body of 1 gram mass which will accelerate it 1 cm/s^2.

E

Earth
This term is used synonymously with soil in an engineering sense. In particular it refers to excavated material when used in engineering structures such as earth dams and earth retaining walls; when used to fill depressions it is known as 'fill' or 'made ground'.

Earth dam
A liquid retaining dam constructed from layers of compacted soil.

Earth fill cofferdam
An enclosure formed by mounds of soil, suitable for use where the head of water is low, the velocity of water flow is small, there is no risk of overtopping and suitable filling material is available.
See also under **Cofferdam**.

Earth pressure
Earth pressure is a push from retained earth (soil) which varies between two extremes, the minimum or active earth pressure (p_a), which is the force from earth tending to overturn a free retaining wall, and the maximum, that is the

passive earth pressure (p_p), which is the resistance of an earth outface to deformation by other forces.

Earth pressure cell
A large non-ferrous flat cell for the *in situ* measurement of soil pressure.

Earth pressure coefficient
See **Active earth pressure**.

Easement
A legal right whereby the owner of one property has certain privileges or conveniences over property belonging to another owner.

Echo sounding test
See under **Geophysical exploration**.

Educator well point
See **Ejector well point**.

Effective diameter
The diameter of a sphere having the same volume as that of an irregular shaped particle.

Effective frontal area
The area of a building or structure normal to the direction of the wind or 'shadow area'.

Effective porosity
The partial volume of the pore space in which the water is free to move per unit total volume of the soil.
In clays the effective porosity may be much smaller than the porosity (n), but in sands the effective porosity and porosity are almost equal in magnitude.

Effective pressure (σ^1)
The pressure in a soil between the points of contact of the soil particles. In a soil system in equilibrium it is equal to the total pressure minus the pore pressure of the water in the pore space. It increases during the consolidation of the soil to a maximum at complete consolidation.

Effective size (D_{10})
The maximum size of the smallest 10% of the particles by mass.

Effective stress (σ^1)
See **Effective pressure**.

Effective stress parameter
These are the angle of internal friction and the unit cohesion of a soil, both with respect to effective stresses.

Effluent seepage
See under **Seepage**.

Egg crate foundation
A reinforced concrete raft foundation stiffened underneath with ground beams forming an egg-crate grid.

Ejector well point
A well point dewatering system whereby a vacuum is created at the well point tip by means of a high speed flow of water through a jet nozzle. Thus the water from the soil is sucked up only a short distance and is then forced up the return pipe.
See also **Well point**.

Elastic compression
The compression of a soil immediately after the application of a load.

Elastic heave
See under **Heave**.

Elasticity modulus
See **Modulus of elasticity**.

Elastic limit
The point on the stress-strain curve of a cohesive elastic soil (or other homogeneous material) beyond which linearity ceases.

Elastomer
A material with rubber-like properties, ease of deformation and rapid and complete recovery.
See also **Grouting**.

Electrical double layer
See **Double layer**.

Electrically conducting paper
See **Conducting paper** and **Electrical seepage analogue**.

Electrical mixer
See **Mixer**.

Electrical piezometer
See under **Piezometer tips**.

Electrical resistivity
See under **Geophysical exploration**.

Electrical seepage analogue
A method of analysing water seepage problems through the soil by passing an electric current through special electrically conducting paper, in which the area to be analysed is cut out to some convenient scale.
An electrical potential is then applied at the boundaries corresponding to the upstream and downstream equipotentials.
With a Wheatstone bridge circuit it is possible to determine the position of any required equipotential.

Electrical vertical settlement system
See **USBR settlement gauge**.

Electrochemical hardening
A special process to improve the strength of clayey soils around piles to enable them to carry higher loads.

Electro-hydraulic system
See **Pressure system.**

Electromagnetic subsurface probing
See **Ground probing radar.**

Electron microscope
A microscope giving an extremely high magnification which enables a researcher to study in detail the characteristics of fine-grained soil particles.

Electro-osmosis
A ground-water lowering process, mainly used in silts to speed up natural drainage and to produce a flow of water away from an excavation. The flow of water is induced by an electrical current flowing from a positive anode to a negative cathode.
See also **Streaming potential.**

Elevating grader
A machine based on a grader but with the addition of a belt elevator to lift the excavated material and deposit it into vehicles travelling alongside.

Eluviation
The movement of soil material, especially colloids, from one place to another within a soil solution or suspension.

Eluvium
Superficial deposits formed of fragmental material from solid deposits which have not been transported by wind or water, but may have moved down hill slopes under the action of gravity when in a waterlogged condition.
See also **Head.**

Embanking
The process of constructing an embankment.

Embankment
A trapezoidal ridge of soil built up in compacted layers to carry a canal, railway or road or to form the containing wall of a reservoir.

Embedment gauge
See **Soil strain gauge.**

Emulsion
A dispersion of one liquid in another.
See also **Grout.**

End bearing pile
See **Bearing pile.**

End isochrome
See under **Isochrome.**

End over end shaker
A mechanical device which can shake a sealed container, such as a glass jar, as well as revolving it vertically through a full circle.

End stop
See **Stop end tube.**

Engineering News formula
An old pile driving formula.
See **Hiley formula**.

Engineering soil map
See **Soil map**.

Enlarged base
An enlargement of the base area of a pile, formed:
(1) with a base larger than the shaft of a pre-formed pile;
(2) *in situ*, by driving a plug of concrete into the surrounding ground; or
(3) *in situ*, by undercutting (under-reaming) the soil at the base of a bored pile.

Envelope of failure
The common tangent to the Mohr circles constructed for a series of similar tests on a soil, under different cell pressures, in the triaxial compression test.
See also **Angle of shearing resistance** and **Angle of internal friction**.

Epoxide
A thermosetting plastic used in soil grouting.

Equilibrium moisture content
The moisture content of a soil in a given environment, at which no moisture movement occurs.

Equipotential lines
Contours of equal water pressure in the soil mass under, through or round a water-retaining structure such as a dam or river bank. The lines are at right angles to the 'flow lines'.
See also **Flow net**.

Equivalent footing analogy
An analogy for replacing the effective bearing area of a pile, or group of piles, by a simple mass foundation.

Erratic condition
A term used in boring or tunnelling to describe vastly changing situations which dictate a variety of approaches to the excavation methods used.
See also **Consistent condition**.

Erythrosine dye
See under **Heave gauge**.

ESP
Electromagnetic subsurface probing.
See also **Ground probing radar**.

Excavation
A hole, trench or cavity made by excavating.

Excavator
A general term for an excavating machine which digs its load by means of a single bucket mounted on, or suspended from, a boom. It differs from other types of excavating plant in that, although mounted on crawler tracks or on wheels so that it can move on a site, or between sites, it is stationary when actually excavating and discharging its load.

Excess pore pressure
The initial extra pore water pressure in a saturated soil when a new higher load is first applied.

Excess pressure
See **Initial consolidation pressure.**

Exchange capacity
A term used to describe the quantity of exchangeable cations in a soil.
See also **Base exchange.**

Expansion
See **Soil expansion.**

Expansion index
The tangent of the slope angle of the straight part of the expansion portion of a voids ratio/pressure line for a soil.

Extension modulus
See **Rubber extension modulus.**

Extension test
A laboratory test to examine axial extension of soil. This represents the heave of soil at the bottom of a deep excavation.

Extensometer
See under **Magnetic probe, Potentiometric, Portable, Tape** and **Transverse extensometers.**

Extractor
See **Pile extractor.**

Extra-sensitive
See **Sensitivity ratio.**

Extremely high liquid limit soil
A soil is said to have an extremely high liquid limit when this value is over 90%.
See also **Liquid limit.**

Extruder
An apparatus for gently extruding undisturbed soil from a sampling tube.
It consists basically of a piston on a long shaft which travels inside the sampling tube. Extruders can be hand, hydraulic or electrically operated and can be fitted with adapters to prepare 38 mm diameter undisturbed samples from U4 tubes.

F

Fabric
A synthetic material in the form of 'cloth' or mesh and used in the construction of pavements, earthworks and French drains.

Face piece
See **Face waling.**

Face shovel
A bucket-type shovel which excavates radially forward and upwards from the machine to which it is mounted. The excavating machine must therefore stand on the floor of the excavation, cutting into a vertical face and moving forwards as work progresses.

Face waling
A waling across the end of a trench supported by the ends of the side walings and which, together with the end strut also acting as a waling, supports the end face of a trench.

Facing wall
The lining usually of reinforced concrete, either precast or cast *in situ*. constructed against the face of an excavation in place of sheeting, and supported by the main 'timbering'. Facing walls are usually left in after construction and are frequently used to receive asphalt water-proofing.

Factor of impact
See **Impact factor**.

Factor of reduction
See **Reduction coefficient**.

Factor of safety (F)
See **Safety factor**.

Facts survey
See under **Preliminary survey**.

Fadum chart
A chart, produced by Fadum in 1941, which gives the influence factors for the vertical stress in a soil at any depth beneath the corner of a rectangular foundation.

Failure wedge
See **Wedge theory**.

Falling head permeameter
See under **Permeameter**.

False bedding
See **Current bedding**.

Fascines
Bundles of twigs about 6 metres long by 250 mm diameter. They are used as a foundation for roads in marshy ground, or as the protection to river banks.
See also **Gabion**.

Fat soil
A highly plastic soil.
See also **Lean soil**.

Fault spring
A spring which occurs when a permeable stratum stops at a geological fault plane against an impermeable rock formation.
See also **Spring, Stratum spring** and **Valley spring**.

Feld's rule
An empirical rule, after Feld, to give the efficiency of a group of piles with respect to their overall bearing capacity.
The value of each pile is reduced by one-sixteenth on account of the effect of the nearest pile in each diagonal or straight row of which the pile in question is a member. The rule is based on using minimum pile centres.
For example a two pile group has an efficiency of 2 at 15/16 = 94% a three pile triangular group an efficiency of 3 at 14/16 = 87% and a group of 5 piles in a square form with the fifth pile at the centre of the group has an efficiency of 4 at 13/16 = 81% and the centre pile at 12/16 = 75% etc.

Fellenius, Wolmar *(1876–1957)*
Chairman of the Swedish Geotechnical Commission which in 1924 first published a report on the application of soil mechanics to practical problems.

Fellenius angles
A series of angles produced by Fellenius to locate the centre of rotation of a critical slip circle relative to a slope in purely cohesive soil.

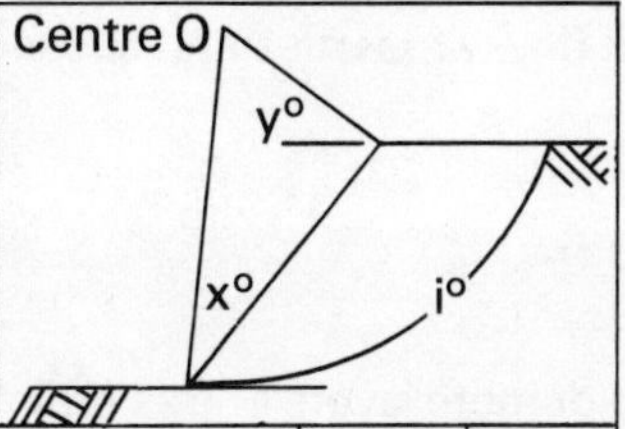

Slope angle i°	*Slope*	*x°*	*y°*
90	Vertical	32	42
60	0.58:1	29	40
45	1:1	28	37
33.8	1.5:1	26	35
26.6	2:1	25	35
18.4	3:1	25	35
11.3	5:1	25	37

Fen
Low lying marshy ground.

Feret triangle
A triangular chart which, if the percentages of the sand, clay and silt fractions of a soil are known, gives at a focal point in the chart the classification designation of the soil.

Field density test
See **Balloon densometer, Density probe, Nuclear densometer** and **Sand pouring cylinder.**

Field drain
See **Agricultural drain.**

Field measurements
See **In situ soil tests.**

Field pumping test
See **Pumping test.**

Field tile
The American name for an agricultural drain.

Fill (or **made ground**)
Suitable inert refuse or excavated soil or rock when dumped for the purpose of filling a depression or raising a site above the natural surface level of the ground.
See also **Overfill** and **Unsuitable fill.**

Fill bulkhead
A bulkhead which after being driven, and the anchorages constructed, has backfill placed behind it such as to form a quay or wharf.
See also **Dredge bulkhead.**

Filler
A finely divided solid which, while taking no part in the chemical reaction in a system, modifies its flow properties and subsequent mechanical behaviour.
See also **Grouting.**

Film camera
See **Borehole camera.**

Filter
This consists of one or more layers of free draining sand/gravel material placed on a less pervious subgrade soil to carry away seepage from, say, a water retaining earth embankment.
See also **Segregation.**

Filter paper drain
Specific size sheets of filter paper that have 6 mm wide parallel slits cut out. They are wrapped round a soil sample, in a drained test, before being placed inside a rubber membrane.

Filter velocity
See **Discharge velocity.**

Filtram
See **Fabric.**

Fine aggregate
See under **Aggregate.**

Fine analysis
An analysis, based on Stoke's law, to determine the distribution of the fine particles in a soil sample.

Fine-grained soil
Soil containing not less than 90% passing a 2 mm BS test sieve.

Fine gravel
Rock fragments ranging in size from 2 mm to 6 mm.

Fineness modulus
The sum, divided by 100, of the percentage masses or volumes of a mineral aggregate retained by each separate standard sieve in succession of a specified series.

Fine sand
Rock fragments ranging in size from 0.06 mm to 0.2 mm.

Fine silt
A natural sediment of rock fragments ranging in size from 0.002 mm to 0.006 mm.

Fine spoil
See under **Colliery spoil**.

Finite element analysis
A versatile tool for numerical analysis which allows complicated boundary shapes and variations in material properties to be introduced into the problem to be solved.
The method can be used for problems involving loads and displacements, seepage, consolidation and dynamic compaction.
The area to be analysed is first divided into a number of geometrically shaped finite elements called sub-regions, each of which is defined by the co-ordinates of the node points. Next, the relationship between quantities such as strain load etc which vary uniformly across the element and their values at the node points are introduced so that the method can be used.

Finite slice method
See **Method of slices**.

Fire clay
A refractory clay.

Firm
Description of a clay having an undrained shear strength of between 50 and 75 kN/m^2.

Fishing
The term used for the operation for recovery of 'lost' tools, or drilling equipment from a borehole.

Fissured
Many stiff clays exist in their natural state with a network of joints or fissures. A large piece of such clay, when dropped, will break into polyhedral fragments. The phenomenon is, in places, accentuated by near-surface weathering due to a seasonal wetting and drying which results in the development of a columnar structure in the clay.

Fixed earth support
A pressure distribution assumed in the design of anchored sheet pile walls where a point of contraflexure is assumed due to the fixed bottom end support of the wall.
See also **Free earth support**.

Fixed head permeameter
See under **Permeameter**.

Flags
See under **Paving flags.**

Flagstone
See **Paving flags.**

Flake thickness gauge
See **Thickness gauge.**

Flakiness sieves
Square vessels with a bottom of woven wire or slotted plate to separate the small flaked soil from the coarse flaked soil. They are manufactured in accordance with British standards.
See also **Sieve.**

Flame structure
Small distorted pockets of sand and silt found in till.

Flapper valve
A one-way flap valve placed between the cutting shoe and barrel of a soil sampler to obtain disturbed samples of sand and gravel.
See also **Core catcher.**

Flexible bulkhead
See **Anchored bulkhead.**

Flexible pavement
See under **Pavement.**

Flexible walls
This type of wall includes thin-stemmed reinforced concrete retaining walls and sheet piled walls.

Flight auger
See **Helical auger.**

Flint clay
A refractory clay.

Float tank
See under **Inverted pendulum.**

Floc
A loosely knit cluster of particles in which face to face contact of the particles is considered unlikely. As the size of flocs increases, the gravitational forces also increase. The flocs thus settle rather rapidly with large quantities of free water entrapped in the soil voids.

Flocculated structure
A description of the arrangement of clay particles where the edge or corner of one platelet tends to be attracted to the flat face of another platelet. The particles thus assume a loose but fairly stable structure that can be maintained as long as the electrical charges on the edges of the platelets remain opposite in sign to those on the faces.

Flood plain
The flat alluvial tract bordering streams and rivers and liable to flooding.

Floury soil
A fine-particled soil which looks like clay when wet but is a powder when dry. It is therefore a silt or rock flour, not a clay.

Flow curve
A graph of the points obtained in the liquid limit test showing number of blows on the horizontal, log scale and water contents on the vertical, arithmetic scale. The point where the curve intersects the 25-blows vertical line is the liquid limit. The flow curve is a straight line.

Flow index
The slope of the flow curve. Since the abscissae are plotted logarithmically it is equal to the difference between the water content at 10 blows and 1 blow or at 100 and 10 blows.

Flowing artesian
See **Artesian water.**

Flowing sand
See **Flow slide** and **Quicksand condition.**

Flowing well
See **Artesian well.**

Flow line paper
See **Conducting paper** and **Electrical seepage analogue.**

Flow lines
Lines in a flow net which show the direction of flow of water through a soil mass (ie near a dam or cofferdam). They intersect the equipotential lines at right angles.
See also **Flow net.**

Flow line tank
See **Seepage tank.**

Flow net
A network of flow lines and equipotential lines intersecting at right angles in a soil mass, such as the cross-section of an earth dam. These lines show the relationship of potential (or head) and flow of water through the soil (near the dam) and they give the following information:

(1) The pore pressure at any point from which the uplift on the dam can be calculated.
(2) The effective pressure on any plane, needed for calculations of dam stability.
(3) The ordinary seepage flow.
(4) The increased pressure due to seepage after rain.

Flow slide
A slide of a liquified mass of loose sand or silt which spreads out to a flat slope after the slide. Those slides which have been analysed have had densities below the critical, that is their void ratios were above the critical voids ratio.

Fluidised bed
See under **Quicksand condition.**

Fluidity
The inverse of viscosity.

Fluvial soils
Soils whose properties are affected predominantly by the action of water to which they have been subjected.
Their common characteristic is roundness of individual particles.

Fly-ash
A product of the burning of powdered coal in a power station. It contains some unburnt coal, the proportion of which does not usually exceed 8%. It is sometimes used as a fill material in civil engineering construction.

Fly-ash grout
See under **Grout**.

Flying shore
A horizontal shore to provide temporary support to two parallel walls, where one or both show signs of failure.

Fog in the hole
Fog which may occur, especially in winter, when concrete is placed in a deep hole.

Foil sampler
See under **Delft sampler**.

Folding wedges
In excavations, wedges used in pairs, overlapping each other and driven in opposite directions in order to hold or force apart two parallel surfaces.

Follower
See **Long dolly**.

Fondu
See **Ciment fondu**.

Foot block
A 'timber' pad used to spread a load from a ground prop or side tree.

Footing
The enlargement of the base of a column or wall for the purpose of transferring the load to the subsoil.
Foundations for dwelling houses or light structures.
See also under **Foundation**.

Footing analogy
See **Equivalent footing analogy**.

Foot piece
See **Foot block**.

Force measuring block
This load-measuring instrument consists of a steel cylinder machined from a solid bar to form four struts. The load on the cell is determined by measuring, with a dial gauge, the average shortening of the struts.

Formation
(1) A series of strata having some common characteristic; also any well-developed stratum or rock mass.
(2) The surface of the subgrade in its final shape after the completion of earthworks.

Former
See **Split former**.

Found
To make a foundation.

Foundation
The supporting base of a structure which transmits the load from the structure to the soil.
The overall movement of the foundation, or the relative movement between separate parts of the structure foundation, must be within the limits that can be tolerated by the proposed structure without affecting its function requirement.
Selection of the correct foundation for a proposed structure on the given soil conditions is complex. The reader is thus given below a cross-reference to other dictionary headings for topics relevant to foundations.

Types of pile
Auger
Bearing
Bored
Box
Brunspile
Cased
Cobi
Composite
Compressed
Compression
Concrete
Driven
Franki
Friction
Fuentes
Group
Guide
H
Holmpress
In situ
Jacked
King
MacArthur
Monotube
Muffler
Palm
Pipe
Pilot
Precast
Precast sectional
Preliminary
Pressure
Prestressed
Raker
Raymond
Rendhex
Rotinoff
Sand
Screw
Sheet
Sheet steel
Shell
Short bored
Silent
Stay
Steel
Tapered
Tension
Test
Thermopile
Timber
Tube
Uncased concrete
Vibro concrete
Wakefield sheet
West
Working

Foundation

Types of footing

Cantilever
Combined
Continuous
Shallow
Spread
Stone
Strip
Tee Beam
Trapezoidal
Unsymmetrical
Wall

Types of foundation

Blob
Compensated
Cone
Crib
Dome
Heavy
Mat
Monolith
Pad
Piled
Rail crib
Sprung
Slab
Stepped

Types of pier

Belled
Bored
Double shaft
Pier shaft
Twin shaft

Types of raft

Cellular
Egg crate
Stiffened

Settlement

Allowable
Differential
Total

Other topics

Bearing capacity
Equivalent footing analogy
Natural foundation frequency
Net loading intensity
Under-reaming

Foundation grillage
See **Crib.**

Foundation mat
See **Raft foundation.**

Foundation modulus
See **Modulus of subgrade reaction.**

Foundation pier
See **Pier.**

Founded
A term applied to a caisson when settled on its bed.

FPS
Federation of Piling Specialists.

Fraction
Soils which have been subjected to a mechanical analysis are described in terms

of their mass percentages of each component, ie the sand fraction, silt fraction, and clay fraction.

Frame
In a trench, any pair of walings on opposite sides of the trench together with the struts that separate them. In a shaft, all the walings and struts at the same level. The word 'frame' is often regarded as including the setting of poling boards supported by these timbers.

Franki pile
An *in situ* concrete pile with an enlarged base. It is formed by pitching a steel driving tube at the required position and placing a charge of aggregate or dry mix concrete in the base of the tube. The aggregate is compacted with a heavy drop hammer working inside the tube to form a plug. Successive blows of the hammer on this plug drive the tube into the ground. On reaching the required level the driving tube is restrained and the plug is forced out from the bottom of the tube by continuous hammering. Dry mix concrete is then added to the tube and hammered out into the bearing stratum until the required resistance is obtained. After the cylindrical reinforcement cage has been lowered into the driving tube, the shaft of the pile is formed by hammering out further charges of dry mix concrete into the ground as the driving tube is gradually withdrawn.

Free earth support
A pressure distribution assumed in the design of anchored sheet pile walls where the wall is considered to be free to rotate about its base.
See also **Fixed earth support.**

Free haul
The maximum distance which excavated material is transported without extra charge. It is based on the specific item in a Bill of Quantities.

Free water surface
See **Water table.**

Freeze
The phenomenom of some piles to increase their load capacity after being driven.
See also **Relaxation.**

Freezing index
The total number of degree days below freezing for a winter. It is an index to estimate the probable depth of frost penetration.
The units of degree days are calculated from the mean night and day air temperatures. For example, if the average temperature is $-5°C$ for two days then this equals 10 degree day units.

Freezing process
The temporary process of *in situ* freezing of water-bearing soil to increase the soil strength as well as to eliminate the passage of water through the soil.

Freezing sampler
See **Cryogenic sampler.**

French drain
Agricultural drains surrounded by filter material such as gravel.

Frictional soil
A clean silt, sand or gravel. A soil whose shearing strength is mainly decided by the friction between particles.

Friction pile
A pile supported by its friction with the surrounding soil.
See also **Pile shoes** and **Precast piles**.

Fringe water
American term for held water just above the water table, which may or may not be permanently held.
See also **Capillary water**.

Frog rammer
A heavy type of power rammer which relies for its compactive effort on the mass of the machine falling back on the soil after being lifted clear of it at each stroke.

Frontage
The length of a site in contact with a road.

Frontage line
See **Building line**.

Frost
Weather during which deposited dew turns to ice.

Frost action
The swelling of soil due to freezing and the subsequent loss of stability on thawing.

Frost boil
The softness of a soil which has thawed after frost heave.

Frosted soil
Soil at sub-freezing or freezing temperature in which the pore ice has not crystallised.
See also **Frozen soils, Supercooled soil** and **Thawed soil**.

Frost heave
Swelling of soil usually upwards due to the expansion of frozen water.

Frozen soil
Soil at negative or freezing temperature in which part of the pore water has frozen.
See also **Frosted soil** and **Supercooled soil**.

Categories of Frozen soil

(1) By time frozen

Description	*Time*
Briefly frozen	Few hours to a few days
Seasonally frozen	One or two seasons
Pereletok	Two seasons
Perenially frozen	3 years to a few decades
Permafrost	Centuries

(2) By iciness

Description	*Ice content*
Low ice	Below 25%
Icy	Between 25% & 50%
High ice	Above 50%

(3) By physical state

Description	*State*
Plastic frozen (high temperature)	Soils with a high unfrozen water content and relatively low compressibility in the frozen state
Hard frozen (low temperature)	Soil rendered firmly cohesive by ice and practically incompressible

Fuentes pile
A concrete sectional pile in which the sections are connected by welding a special sleeve to steel bands on the concrete sections.

Fulbent
A trade name for bentonite.

Fuller formula
The principle of the particle size distribution of a soil, especially in stabilisation, is that the voids of coarser particles should be well filled with finer particles so that a high density may be obtained.
According to Fuller this can be achieved if the particle size distribution of the stabilisation mixture satisfies the following empirical formula:

$$\text{Percentage passing any sieve} = \frac{100\sqrt{\text{aperture size of that sieve}}}{\sqrt{\text{size of the largest particle}}}$$

However, experience has shown that a greater percentage passing the 75-micron sieve is required to obtain sufficient cohesion than that given by the Fuller formula.

Fuller's earth
A clay composed, like bentonite, of montmorillonite.
Used in paints as an extender for its thixotropy.

Full profile gauge
See **Hydrostatic profile gauge**.

G

Gabion
A wire mesh container, of box or mattress shape, filled with stones. The wire mesh is usually galvanised or plastic covered and is taken to the site in a flat pack form where it is assembled and filled with locally found material such as stones. Gabions can be usd as a facing where there is likelihood of soil erosion, as a river training wall or as a simple underwater foundation.
See also **Fascines**.

Gamma ray logging
Strip recording of the intensity of natural radio-activity versus depth, obtained when a suitable detector is lowered down a borehole.

Ganger
The man in charge of a gang of labourers.

Gap graded soil
A soil having a mixture of uniform coarse-sized particles and uniform fine-sized particles, with a gap in gradation between the two sizes.

Gaseous phase
See under **Soil phases**.

Gault
A stiff strata of clay between the layers of green sand in chalk formations.

Geiger counter
An instrument which measures the radio-activity of a substance.

Gel strength
The stress required to break up the gel structure of a bentonite slurry formed by thixotropic build up, under static conditions. It is thus similar to yield point except that the measurement is taken under static conditions.
The usual figure for gel strength is the resistance to flow of the slurry after standing for 10 minutes, measured with a Fann viscometer rotating at three revolutions per minute.

Geodrain
A length of corrugated plastic skin, about 80 mm wide, sheathed in coarse paper which acts as a filter.

Geometry of packing
See **Packing geometry**.

Geomorphology
The study of the topographic features of land.

Geonor cone penetrometer apparatus
A fall cone apparatus designed by the Norwegian Geotechnical Institute, which provides a rapid method of determining the undrained shear strength and the sensitivity of undisturbed and remoulded clays.

Geonor settlement probe
An instrument to monitor long term settlement of soil or ground.
It consists of installing a rod and concentric small bore isolation pipe in a 100 mm diameter borehole. The rod and pipe are carefully pushed into the soil below the level of the borehole and anchored by means of a patent device such as a Borros point. The concentric pipe is then unscrewed from the tip of the rod and withdrawn about 200 mm. The borehole is then backfilled. The settlement of the soil can be monitored by levelling the top of the rod using normal surveying techniques.

Geonor swelling test apparatus
An apparatus which consists of a standard overhead type consolidation frame

with a 10:1 arm ratio. The frame is modified to incorporate a 200 kg load measuring ring secured to a screwjack system to maintain the soil specimen under test at a constant height. The specimen is compacted into a standard consolidation cell of 20 cm^2 cross-sectional area and consolidated at 4.9 kN/m^2.

Geonor vane borer
See **Vane borer**.

Geophone
A seismometer.
See also under **Pulse generator**.

Geophysical exploration
Carrying out a preliminary soil and rock subsurface survey by using special apparatus located on the ground surface, or in a ship or aircraft, to find the change in wave velocity, electrical resistivity, gravity or some other physical variable for which an attempt is made to correlate to the type or properties of the soil and rock.

Geotechnology
The name given to the overall study of geology, soil mechanics and rock mechanics.

Geotextile
See **Fabric**.

Gibbsite
One of the clay minerals, being aluminium silicate $Al_2(OH)_6$ of plate-form structure.

Girder foundation
See under **Crib** and **Strip footing**.

Glacial clay
See **Varved clay**.

Glacial outwash
Material deposited by streams that drained the front of a melting glacier. It is usually composed of gravel, sand and silt.

Glacial soil
Soil which has been transported and deposited by glaciers.

Glacial till
See **Till**.

Glass tank
See **Seepage tank**.

Glotzl pressure cell
A cell for the *in situ* measurement of pressure. The stress around the cell is balanced by an automatically adjusted pneumatic or hydraulic pressure.

Gouda soil sampler
A Dutch designed piston-type soil sampler for obtaining 25 mm diameter by 200 mm long undisturbed soil samples.

Gow caisson
A caisson consisting of telescopic steel cylinders from 1.5 metres to 2.5 metres in length as shaft lining as the excavation proceeds. Each succeeding cylinder is 50 mm less in diameter and is set and driven so that there is about a 300 mm lap with the cylinder above.
The system requires that the ground-water level is lowered by pumping so that it is below the top of the lowest cylinder, assuming that the lowest cylinder is sealed in a cohesive soil. Otherwise, the ground-water level must be kept below the bottom of the excavation.

Grab
A crane fitted with a grabbing bucket which is lowered vertically and digs into, closes round, and picks up the material on to which it has been lowered.
See also **Clamshell bucket**.

Gradation
An American term which refers to the distribution and size of particles in a soil.

Graded filter
Layers of coarse gravel, fine gravel, coarse sand, and fine sand arranged in sequence so that water flowing through the sand does not carry it into the gravel to clog it.

Graded soil
A soil which contains some coarse, fine and medium sizes. It may be well, or badly, graded but is not a uniform soil.

Grade level
See **Formation** (2).

Grader
Earthmoving machine designed primarily for the work of trimming, shaping and finishing the subgrade of pavements and airfield runways.
See also **Elevating grader**.

Grading
(1) Slope of the ground surface, usually by earth moving plant such as graders.
(2) The percentage by mass of different particle sizes in a sample of soil or aggregate, expressed on a grading curve.
(3) The purposeful modification of the proportions of the different particle sizes in an aggregate for concrete or in a soil for an earth dam or other structure, so as to produce the densest or most stable material.

Grading curve
A curve on which the particle size of a sample is plotted on a horizontal, logarithmic scale, and percentages are plotted on a vertical, arithmetic scale. Any point on the curve shows what percentage by mass of particles in the sample is smaller in size than the given point. It is therefore called a particle-size distribution curve.
See also **Particle size distribution chart**.

Grading zone
See **Zone**.

Graft
A narrow, stiff spade especially suitable for digging firm clay.

Graham's salt
See **Sodium hexa meta phosphate.**

Grain
See **Particle.**

Grain shape
See **Particle shape.**

Grain size distribution
See **Particle size distribution.**

Granular soil
A soil which will not form a coherent mass.

Gravel
A natural deposit consisting of subangular to rounded rock fragments usually associated with sandy material. Gravels are roughly classified according to the prevailing size of the contained rock fragments, thus: boulder gravel, cobble gravel, pebble gravel, fine gravel. In mechanical analysis the gravel grade is defined as material between 60 mm diameter and 2 mm BS sieve. A well-graded gravel contains the right proportions of different sizes. A uniform gravel is one with a predominance of a single size.

Gravel auger
An auger specifically designed to bore through gravel and low cohesive soils.

Gravel fraction
The fraction of a soil composed of particles between the sizes of 60 mm and 2 mm.

Gravel-sand cushion
See **Cushion.**

Gravel shoe
See under **Shell auger.**

Gravimeter
An instrument used to measure variations in gravity on or near the surface of the ground.

Gravitational survey
See under **Geophysical exploration.**

Gravitational water
Surface water which percolates downwards towards the ground-water table under the action of gravity.

Gravity abutment
The most common type of bridge abutment. It consists of a central pier supporting the bridge and two wing walls to retain the fill. The three elements rest on a single footing.

Gravity cofferdam
See under **Cofferdam**.

Gravity drainage
The drainage of water from a site or excavation by gravity.

Gravity wall
A retaining wall which depends for its stability entirely from its own mass. No reinforcement is provided except in large mass concrete walls where a nominal amount of 'anti-crack' steel is placed near the exposed surface to prevent cracking with changes in temperature.

Gravity well
See under **Well**.

Grid roller
A towel unit consisting of rolls made up of 38 mm diameter steel bars at 130 mm centres thus giving spaces 92 mm square. Usually there are two rolls, but a third roll can be added to give a greater coverage per pass. The roller is suitable for the initial compaction of all soil types and a typical mass is 5500 kg, but this can be increased to 11 000 kg by the addition of extra mass.

Griffin educator
See **Ejector well point**.

Grillage
See **Crib**.

Grip
A small channel cut into the ground on the uphill side of an excavation to lead rainwater clear of it.

Grit
A coarse-particled sand or sandstone.

Grooving tool
A brass or stainless steel tool manufactured to specific measurements for forming the groove in soil prior to carrying out a Casagrande liquid limit test.

Gross loading intensity
The intensity of vertical loading at the base of a foundation due to all loads above that level.

Ground anchor
A structural member which transits an applied tensile force to 'competent' ground. The shear strength of the surrounding ground is used to resist this tensile force. An anchor may comprise a tension pile or rock bolt etc, but today the most common anchor consists of a high strength steel tendon installed and grouted in at the required inclination to resist the applied load efficiently.

Ground beam
A beam in a substructure transmitting a load to a pile, pad or other foundation.

Ground frame
A 'timber' frame of walings and struts laid 300 mm or so below ground level, used as a guide for the first 'setting' of runners or trench sheeting.

Ground isotherm
See **Isotherm**.

Ground probing radar
A new technique of ground investigation. It consists of an antenna in a wheeled trolley connected by a cable to a control unit, which in turn is connected to a chart recorder and a tape recorder. The information collected is displayed on the chart as it is received while the tape recorder enables more accurate analysis to be made later.
The antenna transmits a core of pulsed microwave radiation into the ground and picks up any axial reflections. Circular cross targets, such as pipes, reflect signals back to the antenna when any part of the cone of radiation hits them, so producing a distinctive trace on the chart. The equipment can also locate such features as permafrost depths.

Ground prop
A prop or puncheon placed between the lowest frame and a foot block on the bottom surface of an excavation and used to support the mass of the 'timbering'.

Ground roughness
The nature of the earth's surface as influenced by small scale obstructions such as trees and buildings.
See also **Topography**.

Ground slag grout
See under **Grout**.

Ground stabilisation
See **Soil stabilisation**.

Ground stabilised cofferdam
See under **Cofferdam**.

Ground-water
Any body of underground water that has collected in a position determined by the porosity and degree of fissuring of rocks and by the underlying impervious strata, thus forming an underground 'reservoir' of water. The water-logged strata below the surface of such water is said to constitute the zone of saturation. The surface of the water is known as the water table, and the level at which the water table occurs is known as the standing water level or ground-water level. Owing to friction encountered by water in its passage through pervious strata, a flowing water table is seldom horizontal.

Ground-water level
See **Water table**.

Ground-water lowering
Lowering the level of the ground-water to make an excavation dry or to cause the sides of the excavation to stand up.

Group index
A method of classifying a soil containing fine-particled material. The higher the group index number, the less suitable is the soil under the AASHO soil classification system.

The group index is calculated from the formula:—

$$(F-35)\,[0.2+0.005\,(LL-40)]+0.01\,(F-15)\,(PI-10)$$

where F is the percentage passing a No 200 sieve expressed as a whole number,
LL is the liquid limit,
PI is the plasticity index.

The group index is always given to the nearest whole number except where its calculated value is negative, when it is given as zero.

Grout
A thin mortar, special chemical or other material used for grouting soil. Grouts can be various types as given below:—

Suspension or particulate grouts
A mixture of ordinary Portland cement and water, although other cementatious materials such as ground slag, fly-ash or pozzolan may replace part or all of the cement.

Solution or non-particulate grouts
Single shot solution grout consists of two solutions, each of one or more chemicals, which are mixed together to give a controllable delayed reaction.
Two shot solution grouts consist of two solutions, each of one or two chemicals, which react almost instantaneously when they come into contact, after being separately injected into the soil.

Hot bitumen grout
A grout often used to inject directly into flowing soil water.

Japanese grout
A special one solution grout which uses only ground-water as a reagent.

Grout curtain
A wall of injected material formed below ground level by the technique of grouting. This is to either form an impermeable barrier or to strengthen the soil.

Grouting
Injecting a cement or chemical grout, under pressure, into the voids of a porous soil to increase its strength as well as to eliminate the passage of water through the soil.
See also **Claquage**, **Clay**, **Grout**, **Open-ended pipe** and **Sleeve grouting**.

GRP
Glassfibre reinforced plastic.

Grubbing
Clearing a proposed construction site of the tree and bush root system.
See also **Clearing** and **Stripping**.

Guide frame
A 'timber' frame erected above ground level to act as a guide for runners or trench sheeting, and as a staging from which they may be driven.

Guide pile
In an excavation supported by sheet piles, a heavy vertical square pile which is

driven close to them and carries the horizontal members (walings) which first guide, and later support the sheet piles. It is strutted to a similar pile on the far side of the dig or to a dumpling or to a raking shore. It carries the full earth pressure from the walings.

Guide runner
In trench work, a runner driven ahead to form a guide for driving intermediate runners.

Guide trench
See **Guide walls**.

Guide walls
Temporary shallow depth walls constructed in soft or unstable soil in advance of the deeper permanent wall.
The two walls are constructed a distance apart equal to the finished permanent wall width plus a suitable working tolerance.
They would, for example, be used in the construction of diaphragm walls.

Gullet
A narrow trench dug to formation level in an earth, or rock, cutting, wide enough to take a track for trucks which remove the soil. The trench is widened as convenient to the full width.

Gumbo
A very fine-particled dark 'plastic' clay deposit, devoid of sand.

Gunite
A pneumatically applied mortar which contains aggregates up to 25 mm in size. It can be used for stabilising poor ground, slopes and tunnels or as reservoir linings.

Guttman process
A chemical consolidation two injection process developed by Professor A Guttman. It differs from the previous Joosten two injection process in that the viscosity of the sodium silicate is reduced before injection by the addition of a suitable salt solution.
This allows the treatment of finer-particled soils to be achieved.
See also **Grouting**.

GWT
Ground water table.

H

Hand auger
See **Post-hole auger**.

Hand boring
A simple and relatively inexpensive method of boring site investigation holes in self-supporting soil, using shell and hand auger, down to a maximum depth of four metres.

Hand carved sample
An undisturbed cohesive soil sample carved from soil in a trial pit or other exposed position.

Hand dog
A short-handled wrench used for making and breaking boring tool rod joints. See also **Bar dog**.

Hand extruder
See under **Extruder**.

Hand mixer
See **Mixer**.

Hand rammer
A wooden or metal block raised and dropped by hand for compacting soil.

Hand vane
See **Torvane**.

Hanger
See **Tie-rod**.

Hanging pendulum
An apparatus to monitor the horizontal movements in dams, abutments etc and foundations.
It consists of a wire anchored at its upper end to a structure under observation. A weight suspended from the lower end is free to move in an oil tank, the oil serving to damp oscillations of the wire. The wire remains vertical but moves with the structure from which it is suspended. Thus readings of movement relative to the wire must be corrected for movement of the anchor.
See also **Inverted pendulum**.

Hansen theory
A general theory developed by J Brinch Hansen in 1953 to account for the various types of possible movement of retaining structures. He has proved by theory and experiments that the earth pressure can be found if a compatible sliding wedge or surface of rupture is used in the calculation.
The surface of rupture may be an arc, a straight line, or a composite curve.

Hard
Refers to a soil which is brittle or very tough.

Hardcore
Hard lumps of stone, brick, furnace slag, old concrete etc, suitable for filling soft ground in a foundation or under a road etc.

Hard frozen soil
See under **Frozen soil**.

Hardpan
The lower horizon of the topsoil in which an accumulation of calcareous or ferruginous cementing material has resulted in the formation of a hard layer between the upper leached horizon of the topsoil and the subsoil. The term is sometimes erroneously applied to rock or other material underlying surface deposits, eg bedrock, or to any material that cannot be classified either as 'rock' or as 'soil'.

Hard water
Water containing calcium or magnesium salts in solution.

Hardy Schulze rule
See **Schulze Hardy rule.**

Hazen, Allan *(1868–1930)*
One of the first persons to make extensive studies of the physical and hydraulic properties of sands and gravels.

Hazen's law
Since the permeabilities of soils vary from 10^{-6} mm/s for clays to 1 mm/s for coarse sand, Hazen's law, based on the effective size of sand, gives an approximation to the permeability based on the particle size. It states that the permeability is approximately equal to the D_{10} size × 1000 in mm/s.

Hazen's uniformity coefficient
The ratio of the maximum particle size of the smallest 60% to the effective size.

H-beam grillage
See under **Crib**.

Head
A heterogeneous mixture of rubble and loamy material derived by weathering from the underlying solid or drift formations. It occurs on high ground and on hill slopes and, in the latter situation, may have travelled a short distance down hill after being converted to a state of sludge by excessive rainfall, melting snow or ice or water from lines of seepage or springs. Under such conditions it tends to accumulate in the concave parts of hill slopes. It is frequently mistaken for topsoil. The term is probably an old quarrying-man's name for overburden and includes material known as hill-wash.

Head boards
Boards at the roof of a heading which are in contact with the ground above.

Header tank
A water container placed at high level to provide a constant head of water for permeability tests.

Heading
A small underground passage, formed to examine the soil strata.

Head tree
A horizontal 'timber' in the roof of a heading, which rests on the side trees and supports the head boards.

Head Wall
A retaining wall at the end of a culvert or drain.

Heave
The rising of the floor of a deep excavation in soil.
The amount of vertical heave and the rate at which it occurs depends on the reduction in vertical stress created by the excavation, on the type and succession of underlying strata and on the ground-water and surface water conditions.
In fairly homogeneous deposits the heave may be separated into two components. First, an elastic type of heave which occurs simultaneously with the progress of excavation, which does not involve migration of pore water

within the deposit. Second, a slow, long-term swelling involving absorption of water from the upper surface or migration from below. This may continue for many years or until reloading by a foundation and supported structure restores the equilibrium.
See also **Frost heave** and **Pile heave**.

Heaved pile
An already driven pile which heaves due to ground displacement resulting from the driving of a subsequent pile of a group.

Heave forces
Those forces producing soil heave and due to frost and to swell of the ground.

Heave gauge
A stud to assist in the monitoring of ground heave due to the formation of an excavation.
The stud is made from four 6 mm thick steel fins welded together to form a crucifix vane 90 mm in diameter and 300 mm in length. A plate 19 mm thick is welded to the top of the fins. The top plate has a tapped hole, 12 mm in diameter, in the centre to aid installation.
To install the heave gauge a 100 mm diameter borehole is drilled vertically into the soil to just below the level of the bottom of the proposed excavation. The heave gauge is lowered on the end of drill rods and forced into the bottom of the borehole until the top plate is flush with the bottom of the borehole. The drill rods are then removed and the borehole back-filled with a dyed bentonite slurry. Red erythrosine dye is usually used so that the borehole can be located during general excavation. By sounding through the bentonite with stainless steel flush jointed 10 mm diameter rods, the top plate is located and its elevation determined using normal surveying techniques.

Heavy foundation
Foundations for heavy industrial plant, such as power stations.

Heavy soil
A soil which is basically clay, and being damper than sand is thus heavier.

Heel
The back of the base of a dam or retaining wall.

Held water
Water held in the soil by surface tension above the standing-water level against the action of gravity.
See also **Aeration zone**.

Helical auger
An auger consisting of a continuous vertical helical thread.

Helmet
A steel 'anvil block' used to protect the head of a timber or precast concrete pile while it is being driven into the soil.

Helve
The handle of an axe, pick or other similar tool.

Herringbone drains
See **Auxiliary drains**.

High air entry piezometer tip
See under **Piezometer tips.**

High ice soil
See under **Frozen soil**.

High liquid limit soil
A soil is said to have a high liquid limit when this value lies between 50% and 70%.
See also **Liquid limit.**

Highly graded
See **Well-graded**.

High plasticity soil
A soil having a plasticity index between twenty and forty.
See also **Plasticity index.**

Hiley formula
A pile driving formula expressed as

$$R = \frac{\eta WH}{s + c/2}$$

where R = the ultimate resistance in tons of the pile to penetration,
s = the final set in inches,
H = the fall on the hammer in inches,
c = the total temporary elastic compression in inches due to pile, helmet, dolly, cushion and ground,
η = the efficiency of the blow obtained as follows:

$$\eta = \frac{W + Pe^2}{W + P}$$

where W is less than P x e

$$\text{or } \eta = \frac{W + Pe^2}{W + P}\left(\frac{W - Pe}{W + P}\right)^2$$

and P = safe load on pile.

Hill-wash
See under **Head**.

Hinged auger
See **Vicksberg auger**.

Hoe
See **Back acter**.

Hoggin
A natural deposit of stony sand or gravel containing a small proportion of clay which is sufficient to hold the mass together without affecting the interlocking properties of the coarser particles.

Hollow stem auger
A continuous helical-type auger having a hollow stem through which sampling

tools can be operated. A removable plug attached to a rod blocks the entry of soil into the stem until the desired depth for a sample is reached.

Holmpress pile
A cast in place reinforced concrete pile especially suitable for use in silty and non-cohesive soils.
The pile is constructed by first driving to the required depth a 400 mm diameter steel tube, fitted with a detachable pointed shoe. The reinforcement cage is then introduced into the pile tube which has internal vanes to position the cage centrally. The tube is then quickly filled with concrete after which the tube only is quickly removed. A further 220 mm external diameter tube, with a detachable pointed shoe, is driven into the wet concrete thus forcing the concrete into the surrounding soil.
The second tube has a further reinforcement cage introduced, concrete is then added and the tube only removed. Further drives into the wet concrete can be made in very poor soil.

Homogeneous strain
See under **Affine strain**.

Honeycomb structure
Some soils are composed of individual particles of silt and flocculated clay arranged in an arching skeleton enclosing large voids. This arrangement is termed a honeycomb structure because of its appearance when viewed through a microscope.

Hookean model
A simple rheological model to represent a perfect elastic response of a material such as soil.
See also **Rheological model**.

Hookes law
This states that up to a certain stress limit, known as the elastic limit, Young's modulus is constant. Beyond this limit the law fails to hold and the soil (or other material) becomes permanently deformed.

Horizontal auger
See **Thrust borer**.

Horizontal shore
See **Flying shore**.

Hot bitumen grout
See under **Grout**.

H-pile
A steel wide-flange column or other section often rolled with a uniform thickness in web and flange.

HRB
American Highways Research Board.

Humidity probe
See **Moisture probe**.

Humus
Dark-brown earthy material, composed of partly decomposed vegetable matter, which forms an important constituent of agricultural topsoil.

Hvorslev, Juul *(born 1895)*
The first authority on the methods of boring and sounding and of obtaining samples of soil.

Hydrated lime stabilisation
See **Lime stabilisation.**

Hydration
A natural process by which water is combined with some other soil substance thus producing certain hydrous minerals such as iron oxide, gypsum and kaolin.

Hydraulic clay cutter
A fast acting, hydraulically operated, boring tool and core sample ejector.
It consists of a 140 mm outside diameter tube having a cutting edge at its lower end. A hydraulically operated 50 mm diameter ram is fixed inside the upper half of the 2000 mm long tube. When the filled tube is brought to the surface of the borehole the ram rapidly pushes the soil core out.

Hydraulic extensometer
See under **Potentiometric extensometer.**

Hydraulic extruder
See **Extruder.**

Hydraulic fill
The placing of fill material in still water, or the process of washing material into place or densifying it with a stream of water.
In general the term hydraulic fill is applied to the complete operation of excavation, transportation, and placing by flowing water. When the hydraulic operation is confined to placing only, it is described as semihydraulic construction. The term sluiced fill is used for material washed into place.

Hydraulic gradient
The potential drop between two points divided by the horizontal distance between the points.
See also **Critical hydraulic gradient** and **Darcy's law.**

Hydraulic head
The head of water (or other fluid) acting at a point in a submerged soil mass. It is expressed by Bernoulli's equation:—

hydraulic head = velocity head + pressure head + elevation head

ie
$$h = \frac{V^2}{2g} + \frac{P}{\gamma w} + Z$$

Hydraulic load cell
A steel-bodied cell with an integral pressure pad and filled with de-aired hydraulic fluid. A pressure sensitive hydraulic transducer fitted to the cell body enables changes in the load acting on the cell to be measured.

Hydraulic overflow settlement cell
A cell for the measurement and control of vertical movement.

Hydraulic

It operates on a U-tube and overflow principle. The cell, a sealed container, is cast into a concrete block, in the structure to be monitored, and is connected by a nylon 'water' tube to a graduated standpipe in a remote read-out location. A second 'drain' tube allows surplus water to flow from the cell, and a third 'air' tube maintains the interior of the cell at atmospheric pressure. Before taking readings compressed air is used to remove water from the air tube and cell. The water tube is then filled with de-aired water by pumping a quantity sufficient to fill the tube and remove any air bubbles. The graduated standpipe which has also been filled is then connected direct to the water tube using a three-way valve on the reading panel. The level of water in the standpipe falls until it reaches equilibrium with the level of an overflow weir in the cell. Thus settlement of the cell is reflected by an equal drop in water level at the standpipe.

Hydraulic piezometer
An instrument to measure and control water pressures in soil.
A porous piezometer tip is installed and sealed at the measuring point and is connected by twin tubes to a remote read-out location. High or low air entry ceramics and tips of differing geometry are used to suit various soil conditions. De-aired water is circulated through the tubes until both the tubes and the tip are completely filled. The pore water pressure at the piezometer tip can then be measured at the read-out location end of either water tube, making a correction for the head difference between the tip and the measuring gauge.

Hydraulic piezometer tip
See under **Piezometer tips**.

Hydraulic pressure cell
See **Pneumatic hydraulic pressure cell**.

Hydraulic settlement gauge
See **Hydraulic overflow settlement cell**.

Hydrocompaction
A form of settlement associated with semi-arid or tropical areas and occasionally with earlier glacial soils. This arises from sudden collapse and occurs principally in sandy or coarse residual soils, following irrigation and inundation.

Hydrofracture
Raising the grouting injection pressure into a soil to induce it to fracture, thus further increasing the grout penetration.

Hydrogen bond
The attraction between the water molecules and the oxygen atoms on the surface of a clay mineral.

Hydrokop trench cutting machine
See **Trench cutting machine**.

Hydrological cycle
The general pattern of water movement at, and near, the earth's surface and mainly involving meteoric water.

Hydrology
The study of the earth's water, including evaporation, ground-water, run off and snow etc.

Hydrometer
An instrument which when floated in any fluid gives a measurement of the specific gravity of the fluid by the level on the stem to which it is submerged. It is used in the wet analysis of soils.

Hydrometry
The use of hydrometers.

Hydrophilic system
A description given to an aqueous colloidal system which disperses readily in water. Nowadays more correctly referred to as macromolecular colloids.

Hydrophobic system
A description given to an aqueous colloidal system which forms stable dispersions in water only with difficulty.

Hydrostatic excess pressure (or **excess pore water pressure**)
The pressure per unit area of soil which exists in the pore water at any time in excess of the normal pore water pressure and due to applied loads. The excess pore water pressure is equal to the total pressure minus the effective pressure plus the normal pore water pressure. Nowadays it is always referred to as excess pore pressure.

Hydrostatic head
See **Hydraulic head**.

Hydrostatic profile gauge
An instrument for the measurement and control of vertical and horizontal movement of soil.
It consists of a probe containing a flexible bladder which is connected by a tube to an electric pressure transducer in the tube drum. The tube and bladder are filled with de-aired anti-freeze solution. A second tube equalises pressures in the probe and in the drum. Changes in level at the probe result in changes in negative pressure at the transducer. The probe is pushed or pulled along an access tube buried either in or beneath the structure to be monitored. Ring magnets are positioned on the access tube during its installation at locations where horizontal movement is to be measured and these actuate a switch in the probe whose position along the tube can be measured.

Hygroscopic coefficient
The moisture, in percentage of its dry mass, that a dry soil will absorb in saturated air at a given temperature.

Hygroscopic moisture
Moisture which is contained in an air-dried soil but evaporates if the soil is dried at 105°C.

I

I-beam grillage
See under **Crib**.

ICE
Institution of Civil Engineers.

Ice phase
See under **Soil phases**.

ICOS-FLEX system
A patented post-tensioned freestanding retaining wall system.

ICOS wall
See **Diaphragm wall**.

Icy soil
See under **Frozen soil**.

Idel Sonde
A probe device to measure horizontal and vertical movement in an earth or rock-fill dam.
The equipment consists of a probe and a transmitter/recorder. 400 mm diameter metal plates with 90 mm diameter central holes are placed in the fill material, at regular intervals, as the dam or embankment is constructed, and a 1 m length of plastic tube 75 mm diameter is fitted through the central hole of each plate, thus allowing the plate to move, relative to the tube, as settlement and deformation takes place. The individual tubes are joined together to form a continuous tube running through the dam or embankment. The amount of movement is observed by measuring the movement of the metal plates. This is done by lowering the probe in vertical holes or drawing the probe along horizontal and inclined holes using a cable. The signal, transmitted through the probe, reflected from the plates, and recorded at the ground surface, locates the plates.

Ignition loss
The loss in mass of organic content of a soil when held over a flame until the soil particles are red hot.

Illite
A crystalline hydrous alumino-silicate mineral of clay.
See also **Kaolinite** and **Smectite**.

Image well
An imaginary recharge well.
Conditions may arise where it is necessary to dewater an excavation located close to a body of water such as a river or canal, where the primary source of water supplying the pervious strata originates from the adjacent body of water. In such cases the edge of the body of water can be considered in plan as an infinite line source of seepage.
To determine the flow to and drawdown caused by the dewatering well, the body of water is replaced by a continuation of the pervious stratum with the same quantity of water as that being pumped from the real well. The image well is located as the mirror image of the real well with respect to the line source.

Immediate settlement
The settlement of a foundation which takes place as the initial construction load is applied.
See also **Consolidation settlement**.

Impact factor
A factor used in design of structural foundations and pavements. The factor is equal to unity when the supported surface load is static.
For moving loads the impact factor depends upon the speed of a vehicle, its vibratory action, the effect of aircraft wing uplift and the roughness characteristics of the pavement (road or runway) surface.

Impermeability factor
The ratio of the amount of rain which runs off a surface to that which falls on it. It is therefore a factor from which the run off can be calculated.
It is as follows:

watertight roof surfaces	0.70 to 0.95
cobblestones	0.40 to 0.50
macadam road	0.25 to 0.60
gravel road	0.15 to 0.30
parks	0.05 to 0.30
woodland	0.01 to 0.20

Impermeable
See **Impervious.**

Impervious
A description of relatively impermeable soils such as clays through which water percolates at about one millionth of the speed with which it passes through gravel.

Impregnation
Allowing a waterproofing liquid to pass through a soil.

Inclined pile
See **Raker pile.**

Inclinometer
An apparatus to measure soil, or rock, movement.
An aluminium access tube, of outside diameter 58 mm, with four internal keyways is grouted approximately vertical in a borehole, or embedded in fill material. The inclinometer probe travels along the length of the tube with its wheels located in one or other pair of keyways depending on the required measuring direction. A strain gauged pendulum within the probe responds to changes in tube alignment. The angle readings taken along the tube can be converted to displacements perpendicular to the tube axis. The readings are recorded on a digital read-out, at a remote location, and this allows automatic summation of displacement values with reference to the initial probe position.
See also **Photographic inclinometer.**

Index of damage
An index, after Don L Leet, where foundations are subjected to dynamic load, such as from an underground explosion.
This index of damage is a limiting displacement of 0.75 mm.

Index properties
Those properties which distinguish one soil from another. They are of two types, the soil particle properties (size, shape, chemical constitution) and

secondly the soil aggregate properties, dry density, moisture content, and consistency limits. The former refer to sands, the latter mainly to clays.
See also **Atterburg limits.**

Indian mattress
See **Gabion.**

Individual footing
A footing that supports a single column.

Inflow
See under **Seepage.**

Influence chart
A chart used to determine the intensity of stress at a point in the ground, at a given depth, when an influence load is placed at various positions on the surface of the foundation.
See also **Fadum chart** and **Newmark chart.**

Influence factor
A calculated value by means of which the vertical pressure at any point, at depth, below a uniformly loaded area, such as foundation slab, can be calculated when using an influence chart.

Influent seepage
See under **Seepage.**

Initial consolidation pressure
The initial pore water pressure in a soil sample about to be tested in an oedometer.

Initial survey
See **Preliminary survey.**

Initial tangent modulus
The slope of the tangent to the stress strain curve, of a soil, at its origin.

Injection
See **Grouting.**

Inside clearance ratio (C_r)
This is given by the formula:—

$$C_r = \frac{D_i - D_c}{D_c} \times 100\%$$

where D_i is the internal diameter of the sampling tube,
D_c is the internal diameter of the crimped tube.
To obtain long undisturbed high quality soil samples it is necessary to reduce the friction between the sample core and the inside of the sampling tube. This is done by crimping the cutting edge so that its internal diameter D_c is slightly smaller than the inside diameter D_i of the sampling tube.
If the ratio C_r exceeds 1%, the sample may expand as it passes into the tube and its strength may be decreased.
See also **Area ratio.**

In situ CBR test
A CBR test carried out directly on the soil at the site of a proposed flexible pavement construction.
The set-up consists of a 45 kN mechanical jack with ball seating connected to a steel joist cantilevered from the back of a heavy vehicle. The load generated by the hand operated jack is indicated on a proving ring and the soil penetration of the 1935 mm^2 piston is measured by a dial gauge registering on a horizontal datum bar assembly.

In situ pile
A pile formed with or without a casing by excavating, driving or boring a hole in the ground and subsequently filling it with plain or reinforced concrete, or precast concrete sections which are grouted in place.

In situ soil tests
Tests made on soil in a borehole, tunnel or trial pit. For example, static or dynamic-penetration tests, vane tests, field permeability tests made by pumping from boreholes, measurement of soil density in place, and vertical or lateral loading tests.

In situ vane borer
See **Vane borer**.

In situ vane test
See under **Vane test**.

Inspection pit
See **Trial pit**.

Inspection shaft
See **Trial pit**.

Intact
A cohesive soil with no apparent fissures is described as intact. This is not affected by the presence of a few inclusions such as fossil shells or other organic material.

Integrator
See under **Moisture probe**.

Intercepting drain
See **Grip**

Interface
The surface or boundary between two soils, or two different materials.

Intergranular cement
The cementing material, such as calcite, iron oxide or silica, which is deposited on the surface of soil particles.
The cementing material may originate from dissolved salts introduced from elsewhere by water flowing through the soil, or it may be the residual product of the disintegration of the soil minerals by leaching.

Intergranular pressure
See **Capillary pressure**.

Interlocking pile
See **Steel sheet piling**.

Intermediate belt
See under **Aeration zone**.

Intermediate failure
A semi-plastic failure, showing shear cones and some bulging of a soil sample in an unconfined or triaxial compression test.
See also **Brittle** and **Plastic failure**.

Intermediate liquid limit soil
A soil is said to have an intermediate liquid limit when this value lies between 35% and 50%.
See also **Liquid limit**.

Internal clearance ratio
See **Inside clearance ratio**.

Internal friction
See **Angle of internal friction**.

Interpolation
Inferring the position of a point between two known points by assuming that the variation between them is linear.

Inverted pendulum
An apparatus to monitor horizontal movement in dams, abutments etc, and foundations.
The inverted pendulum is installed to provide a fixed datum from which structural movements can be monitored. It consists of a wire anchored in stable ground beneath the structure, with a float fixed to its upper end. The float which is free to move in a water tank, tensions the wire and keeps it vertical. Displacements relative to the wire can then be measured.
See also **Hanging pendulum**.

Invert level
The lowest level of the floor of a channel, culvert, drain etc.

Inward seepage
See under **Seepage**.

Ion
A positively charged particle.
See also **Anion**, **Base exchange**, **Cation** and **Co-ions**.

Ionic bond
The force of attraction when an atom is positively charged by the loss of one or more electrons is attracted to atoms negatively charged by the gain of electrons.
See also **Van der Waal forces**.

I-pile
See **H-pile**

Iron pan
Hardpan, sometimes impervious, cemented by iron oxides.

Irregular particle
See **Particle shape.**

Island caisson
See **Sand island caisson.**

Isochrome
A time curve showing the proportion of the pore water pressure and effective stress in a clay soil between the beginning (t=o) and end of consolidation (t = ∞) for a certain load increment at a certain time t.
The isochrome for time t = o is termed the zero isochrome and for time equal to infinity is termed the end isochrome.

Isoelectric point
The point at which the positive and negative charges of a soil particle balance.

Isogeotherm
Depths of equal temperature in the soil.

Isohyet
A line joining points of equal rainfall.

Isohyetal map
A map which shows relative amounts of rainfall by the use of isohyet lines.

Isolated footing
A footing that supports a single column.

Isoptera
See **Termite.**

Isoshear
Having the same shear strength.
A typical example is the lines of isoshear forming pressure bulbs under a loaded foundation.

Isotache
A graph relating intergranular pressure to void ratio for a specific constant consolidation speed.

Isotherm
Contours of equal temperature as plotted on, or under, the surface of the earth.

Isothermal
Having the same temperature.

Iwan auger
See **Post-hole auger.**

J

Jacked pile
A pile forced into the ground by jacking.
See also **Silent piling.**

Jacking pressure
The pressure required to force a pile into the ground.

Jack load cell
See **Pneumatic/Hydraulic pressure cell.**

Japanese grout
See under **Grout.**

Jarusov rule
Ions with the highest bonding energy preferentially occupy sites on the surface which have the greatest mean free bonding energy.

Jetting hole
See **Pile water jet.**

Joint
A crack, sometimes found in clay, formed as a result of desiccation after deposition of the soil.

Joosten process
Injection of two separate solutions through pipes driven or jetted into the soil. The solutions injected are calcium chloride after sodium silicate. The two solutions meet in the ground and react to form a gel which strengthens the soil and makes it watertight. The method is suitable only for sands and gravels, but in them it forms a solid cylinder of about 600 mm diameter round each injection tube.

Junction growth
The increase in the cross-section area of a material such as a soil sample, with applied shear force.

Juul Hvorslev
See **Hvorslev, Juul.**

K

Kaolinite
A two-layer clay mineral being a crystalline hydrous alumino-silicate.
See also **Illite** and **Smectite.**

Karl Terzaghi
See **Terzaghi, Karl.**

Keel
A longitudinal protrusion on the underside of a retaining wall foundation to prevent the wall sliding away from its retained material.

Kelvin model
A simple composite rheological model to represent a viscoelastic response of soil.
See also **Rheological model.**

Kentledge (or **cantledge**)
Loading to provide a reaction over a jack, to push down a plate in the plate

bearing test, or to test a caisson or bearing pile. It may be scrap metal, large stones, tanks filled with water or any other convenient material.

Kepes model
A rheological model to represent a general plastic solid in which the magnitude of the yield point is a function of the strain.
See also **Rheological model**.

Kerf area
See **Area ratio**.

Kern
The locus of the most eccentric points of application of the resultant vertical load corresponding to compression over the base of a foundation slab.

Key
A nib formed longitudinally below the base of a thin retaining wall to prevent the wall from sliding forwards. It is usually an extension vertically downwards of the wall stem although the nib may protrude below the heel or toe of the base of the wall.

Kicking piece
A length of 'timber' spoked to a waling to take the thrust from the end of a strut which is not at right angles to the waling.

King pile
In a wide, strutted sheet pile excavation, a long pile driven at the strut spacing in the centre of the trench before it is dug. King piles thus halve the length of the main 'timbers' from side to side of the excavation, since the main 'timbers' from each side bear on the king piles.

Knotts clay
Oxford clay found around Peterborough. It contains carbonaceous matter and is used in the manufacture of Fletton bricks.

Kozeny-Carman equation
This equation relates the coefficient of permeability to physical properties of the system such as porosity and the surface area of the particles.

$$k = \frac{1}{k_o S^2} \times \frac{\gamma}{\mu} \times \frac{e^3}{1+e}$$

where k_o is a constant depending on the pore shape and ratio of length of flow path to soil stratum thickness,
S is specific surface area of particles,
γ is density of permeant,
μ is permeant viscosity,
and e is voids ratio.
See also **Darcy's law** and **Loudon's formula**.

L

Laboratory vane test
A bench-mounted apparatus, designed by the Transport and Road Research Laboratory, to determine the shear strength of low strength cohesive soils

whilst the sample is still in its sample tube.
The apparatus comprises a hand-operated frame with a base plate capable of acceptable sample moulds and tubes. Load is applied through the 12.7 × 12.7 mm crucifix vane by means of any one of four calibrated springs. The motorised attachment gives a shearing rate of 10 degrees per minute.
See also **Vane test**.

Lacing
A vertical 'timber' spiked to the sides of struts or walings and tying them together to carry the mass of the lower frames as excavation advances.

Lacustrine soil
Soil deposited in freshwater lakes.

Ladder trencher
See **Trench cutting machine**.

Lagging
Horizontal boards wedged against a recently cut soil face.

Lake marl
See **Boglime**.

Laminated
Some cohesive soils exhibit laminations which are more or less parallel with one another and with the bedding planes. Varved clays are a special class of laminated clays.

Land drain
See **Agricultural drain**.

Landslide
See **Landslip**.

Landslip
A sliding down of the soil on a slope because of an increase of load, decrease in shear strength or the removal of support at the foot due to cutting a railway, road or canal. Clays are particularly liable to slips.

Laser beam
A technique for measuring lateral movement of structures.
The laser beam is set up from one reference point and sighted onto a second distant reference point. The relative movement of predetermined points on the structure can then be measured as offsets from the laser beam reference line.

Lateral earth pressures
The lateral forces interacting between an earth retaining structure and the retained earth mass.
See also **Active earth pressure** and **Passive earth pressure**.

Lateral pile load test
An *in situ* test to estimate the lateral resistance of a driven pile.

Lateral strain indicator
A device to mechanically or electrically measure the lateral strain of a cylindrical soil sample.

Lateritic soil
A reddish brown residual sandy or gravelly soil, often hard and compact, formed by leaching and weathering, under tropical conditions. The resultant material is rich in aluminium and iron, and primarily composed of the oxides of these metals. The harder varieties are used for roadmaking.

Lathe
A lathe for preparing cylindrical soil samples from rough samples. It can be either hand or electrically operated.

Law of mass action
If an insoluble product forms as the result of exchange the reaction proceeds to completion.

Leach
To remove salts from soil, or minerals from an ore, by passing water through it.

Leaf clay
See under **Varved clay**.

Lean soil
A slightly plastic soil.
See also **Fat soil**.

LECA
Light expanded clay aggregate.

Leg
A support in an underpinning forming part of the permanent work.

Legend
See **Soil legend**.

Leighton Buzzard sand
A clean, white, closely-graded sand, used as a 'standard sand' in tests.

LI
Symbol for liquidity index.

Lifting dog
A simple lifting tool having a twice bent steel shank fitted with a lifting eye and used for the rapid insertion or extraction of boring tools. The collar of the boring tool locks behind the upright stem of the tool lifter.

Light beam
See **Laser beam**.

Light soil
Coarse-particled soil with a very low silt and clay content and incoherent single-particled structure.

Lignin
A waste product resulting from paper manufacture and used in chemical grouting.
The lignin, combined with a diachromate, forms a soluble chrome-lignin powder which gels to a dark brown substance of rubbery consistency.

Lime soil
A soil containing calcium carbonate as the predominant part. The proportion is usually under 50% but may rise to over 75%.

Lime stabilisation
Mixing hydrated lime into a swelling clay to reduce its plasticity and swelling characteristic. Also adding lime to clays and silts to improve their workability, if their moisture contents are above the optimum for compaction.

Limit of liquidity
See **Liquid limit**.

Limit of plasticity
See **Plasticity index**.

Limits of consistency
See **Consistency limits**.

Limnoria
A marine organism which attacks timber piles in brackish or salt water.

Linear shrinkage (LS)
Linear shrinkage indicates the plastic properties of a soil with a low clay content.

Line of seepage
The steepest slope of the water table.

Line of thrust
A line containing the resultant thrusts on the back of a retaining wall or timbered trench.

Liner
(1) A thin-walled tube placed inside a reamed out split spoon soil sampler. This enables the liner and the undisturbed soil sample that it contains to be removed from the spoon. After waxing the ends of the sample and capping the ends of the liner the whole can be sent to the laboratory for testing.
(2) A 'timber' driven between the ends of opposing members of a frame to lock them in position and spiked to the member against which it rests.

Lip
A short length of 'timber' fixed and spiked to the top of a strut, and projecting sufficiently beyond its end so as to rest on a waling. It supports the mass of the strut while wedges are being driven.

Lipping block
See **Lip**.

Liquefaction
A saturated sand which has a void ratio higher than the critical void ratio, that is a loose sand, is liable to liquefaction when it is shocked by earth tremors or by more water flowing into it or by sudden loading. The sand becomes temporarily quick and a flow slide is caused.

Liquidity index (LI)
The LI of a cohesive soil is defined as:

$$\frac{\text{natural moisture content} - \text{plastic limit}}{\text{liquid limit} - \text{plastic limit}} \times 100\%$$

It is a measure of the consistency of the soil. Soft clays have a LI approximating to 100%, while stiff clays have a LI which approximates to zero and may even be negative.
See also **Atterburg limits**.

Liquid limit (LL)
The minimum moisture cntent at the point between the liquid and the plastic states of a clay.
See also **Atterburg limits**.

Liquid limit device
An apparatus, after A Casagrande, for the determination of the liquid limit of a soil.
See also **Cone penetrometer apparatus**.

Liquid phase
See under **Soil phases**.

Live load
A movable load on a structure.

LL
Symbol for liquid limit

Load cell
An enclosed cell incorporating strain gauges which, when incorporated in an electrical 'bridge' circuit, accurately measures the load applied to a soil sample under test.
See also **Disc load cell, Photoelastic load indicator** and **Vibrating wire load cell**.

Load factor
Multiplying the design load by a numerical factor to provide for a factor of safety on the completed design.

Loaded filter
A graded filter at the foot of an earth dam or elsewhere, which stabilises the toe of the dam by its mass and permeability, since no water can exist in it under pressure.
See also under **Earth dam**.

Loading caps
Special aluminium or Perspex top end pieces which transmit the load from the loading ram to the soil sample in a compression test such as the triaxial test.
For undrained tests a plan disc of Perspex with a central coned seating in its upper surface can be used. This seating takes a stainless steel ball which registers in a conical seating in the lower end of the loading ram. This ensures that the load is applied centrally and allows freedom of movement to the top of the sample under test. Where the deformation characteristics of the soil sample

are important a halved steel ball and flat ended ram can be used.
For consolidated undrained or drained tests a cap with a tubular projection making a loose sliding fit on the ram is used.
When drainage from the top of the soil sample under test is required, a 1 mm internal diameter polythene tube is fitted to the loading cap which has an annular groove on its underside.

Load ring
See **Proving ring**.

Loam
A general term largely used to refer to soft deposits consisting of a mixture of sand, silt and clay in varying but approximately equal amounts.

Loess
A yellowish light brown uniform, cohesive, wind blown, porous, but coherent deposit of very fine material ranging in size between 0.01 mm to 0.05 mm diameter.

Log
See **Borehole log**.

London clay
A stiff fissured clay found beneath London and the surrounding districts.

Long dolly
Generally a hardwood 'pile' fitted into the top of a pile driving helmet so that the top of the structural pile can be driven into the ground below the level of the pile driving frame. Both the helmet and dolly are recovered when the structural pile has been driven to its correct level.

Loose ground
Loose non-cohesive material which can be easily excavated with a shovel; in quantitative terms, a material may be regarded as loose if its relative compaction is less than 90%. In certain applications it may be desirable to ascertain how the dry density compares with the 'critical density' of the material.

Loss of ground
In excavating soft ground the volume of silt, clay or sand dug out is greater than the volume of the excavation because material flows into it even when no boil occurs. This ground is lost from outside the excavation and may cause settlement of neighbouring buildings. To avoid this, excavation must be done in sheet-piled cofferdams.

Loudon's formula
An empirical formula relating permeability to the geometrical properties of soil particles:

$$\mathrm{Log}_{10}(kS^2) = a + bn$$

where k = coefficient of permeability,
S = specific surface,
n = porosity,
and a and b = constants.

For permeability at 10°C, the values of the constants a and b are 1.365 and 5.15 respectively.
See also **Kozeny–Carman equation.**

Low air entry piezometer tip
See under **Piezometer tips.**

Lower capillary fringe
See under **Aeration zone.**

Low ice soil
See under **Frozen soil.**

Low liquid limit soil
A soil is said to have a low liquid limit when this value is below 35%.
See also **Liquid limit.**

Low plasticity soil
A soil having a plasticity index between five and ten.
See also **Plasticity index.**

LS
Symbol for linear shrinkage.

Lugeon unit
The flow of water in litres per minute per metre of hole.
In cases where *in situ* permeability tests are carried out to investigate the applicability of grouting methods the results of the tests are expressed in Lugeon units. The tests are conducted with measurements of water flow over a borehole length of 5 m usually under an applied pressure difference of 1000 kN/m^2.

M

MacArthur pile
A composite pile which has an uncased concrete top section and a timber bottom section.
It is formed by driving a steel casing and core, withdrawing the core, driving a timber section through the casing, placing a batch of concrete, replacing the core on the concrete to form a plug, withdrawing the core, placing a further batch of concrete and withdrawing the casing while the mass of the core rests on the concrete. This procedure is repeated until the correct height of concrete is reached.

Mackintosh probe
A solid spear-shaped probe, 127 mm long by 29 mm maximum diameter, which screws onto the lower end of a string of rods. It is then forced into the soil under a mass of 2.4 kg falling 610 mm sliding over the drilling rods. The rate of penetration is an indication of the soil strength.

Mackintosh prospector
A portable sampling kit, contained in a carrying box, for boring and prospecting down to a depth of 15 m.

Macromolecular colloids
See under **Hydrophilic system**.

Made ground
See **Fill**.

Magnetic probe extensometer
An apparatus for measuring soil and rock movement. Ring magnets sliding on a central access tube are fixed in the ground at locations where movement is to be monitored. A probe incorporating switches travels within the access tube and senses the positions of the magnets outside the tube.
See also **Potentiometric** and **Tape extensometers**.

Magnetic survey
See under **Geophysical exploration**.

Magnetometer
An instrument used for measuring the intensity and direction of the earth's magnetic field.

Maintained load test
A pile loading test in which each increment of load is held constant either for a defined period of time or until the rate of movement of a pile falls to a specified value.

Mandrel
A revolving shank.

Manesty still
A boiling chamber for the continuous production of distilled water.

Man lock
An air lock through which men, not materials, pass.
See also **Air lock**.

Manometer board
A wall-mounted board having a series of manometers fixed to it for use in permeameter and model dam tests.

Margin of safety
See **Safety factor**.

Marine soil
Soil deposited through sea water.

Marl
Soil (or rock) containing a quantity of calcium or magnesium carbonate less than 50% of the whole by mass.

Marsh
Ground liable to frequent flooding but supporting vegetation; often, but not necessarily, consisting of alluvium.

Marsh funnel
A simple cone-shaped site instrument for measuring the viscosity of slurry.
The test consists of covering the lower end of the funnel with the operator's

finger, pouring in the slurry through the top mesh to the base of the mesh screen, and then immediately letting the slurry flow out into a special container and noting the time taken to fill the container.
See also **Mecasol cone** and **Prepakt cone.**

Marsh-gas (CH_4)
Methane gas.

Marston theory
A basic theory for the determination of loads on buried pipelines after Dean Marston of Iowa State University.
The marston load formula is

$$W = C\,w\,B^2$$

where W is the load on the pipeline,
w is the weight of the backfill,
B is the breadth between the friction planes,
and C is a coefficient and is a function of H/B and the friction characteristics of the backfill.

Masonary footing
See **Dimension stone footing** and **Rubble stone footing.**

Mass action
See **Law of mass action.**

Materials lock
An air lock through which excavated soil and other material is passed out, and construction material passed in.
See also **Air lock.**

Mat foundation
See **Raft foundation.**

Matrix suction
The negative pore water pressure in a soil at the air/ground water level interface. In clean sands this suction seldom exceeds pF = 1.7, but in oven-dried clays where the capillaries are much smaller suctions up to pF = 7 are obtained.

Mattress
See **Gabion.**

Maximum dry density
The dry density obtained by a stated amount of compaction of a soil at the optimum moisture content.

Maximum safe bearing capacity
The maximum value of contact pressure to which soil can be subjected without risk of shear failure, ie the ultimate bearing capacity of the soil divided by a suitable factor of safety.

Maximum thaw depth
See **Thaw depth.**

Maxwell model
A simple composite rheological model to represent an inverse viscoelastic response of soil.
See also **Rheological model**.

MCV
Moisture condition value.
See also **Moisture condition value test**.

Measuring block
See **Force measuring block**.

Mebra drain
A vertical band drain, 95 mm wide, invented in Holland to dissipate pore water pressure.
It consists of a central flexible plastic strip incorporating about 40 longitudinal drainage channels surrounded by a filter of paper treated against rot.

Mecasol cone
A simple cone-shaped instrument for measuring the viscosity of slurry at site. It is slightly different in shape from the Marsh funnel although used in a similar manner.
See also **Marsh funnel** and **Prepakt cone**.

Mechanical analysis
The determination of the proportions of the different particle sizes in a given soil sample to help in classification of the soil. Particles larger than 0.06 mm can be sieved. Smaller particle diameters are estimated from their settling velocities in water, by wet analysis.

Mechanical weathering
The breakdown of a mass of rock into smaller size particles without chemical change. Thus the type of soil derived depends on the type of the disintegrated parent rock.

Medical lock
See **Air lock**.

Medium gravel
Rock fragments ranging in size from 6 mm to 20 mm.

Medium-particled soil
Soil containing not less than 90% passing a 20 mm BS test sieve.

Medium pit
See under **Pit**.

Medium plasticity soil
A soil having a plasticity index between ten and twenty.
See also **Plasticity index**.

Medium sand
Rock fragments ranging in size from 0.2 mm to 0.6 mm.

Medium silt
A natural sediment of rock fragments ranging in size from 0.006 mm to 0.02 mm.

Medium trench
See under **Trench**.

Membrane
(1) Thin rubber or plastic sleeve used to contain a soil sample in a triaxial compression test.
(2) A thin waterproof layer of plastic or rubber placed under, in or around a foundation to prevent the ingress of ground-water.

Membrane apparatus
See under **Membrane stretcher** and **Pressure membrane apparatus**.

Membrane correction
A correction to be applied to the results of triaxial compression tests to allow for the restraint imposed on the soil sample by its enclosing rubber or plastic membrane.
The correction to be applied to the measured compression strength, due to the effect of a rubber membrane is given by:

$$\frac{\pi DM \varepsilon}{A}$$

where D is the initial diameter of the test sample of soil,
M is the compression modulus of the membrane per unit width,
and A is the corrected cross-sectional area of the sample at axial strain ε.
See also **Rubber extension modulus**.

Membrane cut-off
See **Cut off wall**.

Membrane stretcher
A thin-walled tube with a flexible suction tube fixed centrally at right angles to its side. Used to aid the placing of a rubber membrane over a soil sample before testing in a triaxial compression cell.

Menard pressure meter
An instrument to determine the *in situ* strength and deformation modulus of soils and weak rock.
It consists essentially of a probe which is placed in a borehole at the required depth and then expanded. The probe consists of a metal cylindrical body over which are fitted three cylinders. A rubber membrane covers the cylinders and is clamped between them to give three independent cells. The cells are inflated with water and a common gas pressure is applied to provide a radial stress. The deformation of the central cell can then be measured volumetrically.

Meniscus
The curved upper crescent-shaped surface of a liquid column in a capillary tube.

Mesh
See under **Fabric**.

Meteoric water
Water derived from the atmosphere by precipitation.

Methane (CH_4)
Known as marsh-gas in marshes or firedamp in mines.

Method of slices
A method of calculating the stability of a soil slope. The wedge of soil is divided into vertical slices of equal width but different heights, and therefore of different frictional resistance to disturbance. The equilibrium of the whole wedge is calculated by summing the effects on all the slices.

Methylene blue test
A test to determine the cation exchange capacity of a slurry.

Meyerhof's formula
A formula which gives the bearing capacity P_u of a driven pile in a soil possessing both cohesion and friction where

$$P_u = A_s (c_a + K_s \gamma D \tan \delta) + A_b \left(cN_c + K_s \gamma DN_q + \frac{B}{2} \gamma N_\gamma \right)$$

where A_b = area of the pile base,
A_s = area of the pile shaft,
B = breath of the pile,
c = cohesion of soil,
c_a = adhesion per unit area,
D = depth of pile,
K_s = the coefficient of earth pressure on the pile shaft within the failure zone, varying from ½ for loose soil to about unity for dense soil
N_c, N_q & N_γ = bearing capacity factors that are dependant on the angle of shearing resistance ϕ and the embedment ratio D/B,
δ = angle of friction of the soil on the pile shaft,
and γ = density of the soil.

Micellar solution
A term often used to describe the solution in the double layer of a colloidal particle.

Mineral soil
A soil composed mainly of inorganic matter, in contrast to a soil composed largely of organic matter.

Mini earth pressure cell
A miniature diaphragm cell 26 mm outside diameter by 6 mm thick providing an electrical output from two internal strain gauges when used in conjunction with a strain 'bridge'.

Mining subsidence
Surface subsidence due to underground workings.

Mini piezometer tip
See under **Piezometer tips**.

MIT
Massachusetts Institute of Technology.

Mixer
A laboratory unit for preparing theoretical samples of soils to specific gradings and moisture contents. The unit, which can be either hand or electrically driven,

consists of a small bowl, or pan, and internal paddles which rotate in opposite directions to ensure a thorough mixing of the ingredients.

Mix in place
Soil stabilisation in which the soil is mixed by a travel mixer without being removed from the ground.

Mobile density logger
See **Road logger**.

Modified loess
This is formed from loess by processes of immersion, saturation, erosion or decomposition. All of these reduce the porosity and may alter the mineral composition.

Modulus of compressibility
The ratio of the pressure in the soil mass to the volume change caused by the pressure.

Modulus of elasticity
See **Young's modulus**.

Modulus of extension
See **Rubber extension modulus**.

Modulus of foundation
See **Modulus of subgrade reaction**.

Modulus of subgrade reaction

The value K in the empirical equation:

$$\text{soil pressure} = K \times \text{settlement}.$$

This equation is not always reliable but may be a useful working assumption up to a limiting pressure.

Modulus of ventilation
See **Ventilation modulus**.

Modulus of volume change (m_v)
The coefficient of compressibility (a_v) divided by (1 + initial voids ratio).

Mohr envelope
See **Envelope of failure**.

Mohr's circle of stress
A graphical construction which enables the stresses in a cross-section oriented at any direction to be easily determined if the principal stresses are known. It is commonly used for determining stresses in two directions and can be used for determining three-dimensional stresses.

Moisture condition value test
A test developed by the Transport and Road Research laboratory for the rapid measurement of a moisture condition value (MCV) of earthwork material.
The test is intended as a means of assessing the suitability of earthwork

materials for use in embankments, while avoiding the measurement of moisture content with its associated delays.
The performance of earthmoving plant can be related to MCV, and a relationship between undrained shear strength of remoulded soil and MCV also exists. The test basically consists of determining the compactive effort necessary, in terms of the number of blows of a rammer, to fully compact a sample of soil. At any given moisture content a comparison of the densities produced by two compactive efforts will indicate whether the lower compactive effort was sufficient to produce full compaction.

Moisture content (m)
The mass of water in a soil sample divided by the mass of the solids and multiplied by 100.

Moisture index
See **Atterberg limits**.

Moisture meter
See **Speedy moisture meter**.

Moisture probe
A pressure-tight stainless steel cylinder containing a radio-active source and a sensing device which is pushed into the soil. When connected to a surface integrator the *in situ* moisture content of the soil can be found.
See also **Density probe**.

Mole drain
A drainage slot cut into a clay soil by drawing a mole plough through it.

Mole plough
This consists of a cylindrical cutting shoe fixed at the bottom of a vertical knife-edged tine. The cutting shoe can be adjusted to a depth of between 300 and 600 mm below ground level. In front of the vertical mole tine is a cutting disc which cuts to a depth of about 100 mm to lessen the resistance against the mole tine. The effect of the tool is to leave a round drainage tunnel 50 mm in diameter connected to the surface throughout by a narrow drainage slit. Gravel can also be made to trickle down behind the blade to assist in maintaining the capacity of the drain.

Moling
Laying drains with the aid of a mole plough.

Moment
The turning effect of a force about a point.

Monolith
An open caisson of heavy mass concrete or masonry construction, containing one or more wells for excavation.

Monolithic walls
An umbrella term to cover retaining walls of the cantilever, counterfort, gravity, and semi-gravity type.

Monotube pile
A fluted, tapered concrete shell pile which is heavy enough to be driven into the ground without the aid of a mandrel.

Montmorillonite
A dioctahedral-shaped clay mineral having a typical chemical formula:

$$n[(Al_{1.67}Mg_{0.33})Si_4O_{10}(OH)_2]$$

Montronite
A clay mineral similar to montmorillonite.

Morainal deposit
See **Till.**

Mottling
A term used to describe the overall colour effect of a soil where two or three colours occur in patches.

Muck
Waste rock or excavated soil.

Muck lock
See **Materials lock.**

Muck shifting
Earth-moving.

Mud
Silt or clay in an almost fluid condition. Areas of mud supporting no vegetation and at times covered by floods are known as mud-flats.
See also **Drilling mud.**

Mudcake
The formation of a relatively impervious lining on the walls of a borehole when drilling with the aid of drilling mud.

Mud-flats
See under **Mud.**

Mud flow
See **Flow slide.**

Mud jacking
Boring a hole through a concrete road or foundation slab which has sunk and connecting it to a mud jack by flexible pipe. The mud jack is a lorry on which is a mixer for water-cement-soil slurry and a pump which forces the mix under the slab and raises it.

Mud line
See **Dredge level.**

Mud pump
See **Sludge pump.**

Mud run
A flow of mud into an excavation.

Mudstone
An indurated, structureless deposit of clay material, often interbedded with clay shales but distinguishable from them in being unlaminated and breaking, under

the hammer, with irregular or conchoidal fracture. Mudstone is common in more fine-particled parts of the coal measures. In Midland and North of England colliery districts it is termed bind.

Mud wave
A slide along a cylindrical surface which produces lateral and upward displacements of the underlying soft clay.

Muffler piling
A system of driving a panel of several steel sheet piles quietly into the ground using a double-acting air-driven hammer.
This is achieved by using a pair of acoustic shrouds which act in combination. The steel sheet piles are pitched between a pair of conventional gates which have first been surrounded by a horizontal acoustic enclosure; a second vertical shroud, which can both telescope and travel sideways, is then lowered on to the first enclosure. This permits each pile in the panel to be driven in turn first down to the level of the gate, then down to the required height above ground.
See also **Silent piling**.

Multiple piezometer
This consists of several ceramic filters fastened together by lengths of pipe. Each filter is connected to a separate small-bore plastic tube which is brought to the surface inside the pipe. This piezometer is suitable for installation in a borehole, provided each filter is surrounded by sand and is separated from its neighbour by an impermeable backfill.

Multiple point extensometer
An apparatus to monitor long-term movement in soil. It consists of several steel wires or cables, separately attached to various anchor zones and passing through guide tubes to measuring points at the surface where the movement of a point on a cable is observed relative to a fixed reference point, the tension in the wires being held constant.
The apparatus is set up in a borehole which can be vertical, inclined or horizontal.

Multi-stage well point system
See under **Ring system**.

Muscovite
A dioctahedral shaped mineral having the chemical formula

$$KAl_2\,(Si_3AlO_{10})\,(OH)_2$$

N

Nail spike
See **Brob**.

Natural foundation frequency
The frequency of free vibration of a foundation-soil system. It is important that this frequency should be appreciably different from that of any machines which the foundation carries, to avoid resonance.

Natural gas
Gas which flows from the ground and can be used as a fuel. It is mainly Methane CH_4.

Natural moisture content (m)
The natural or *in situ* moisture content of a soil is defined as the mass of water contained in the pore space, expressed as a percentage of the dry mass of solid matter present in the soil.

Navvy
An old name for a skilled labourer who dug foundations, trenches etc.

Navvy pick
A double-pointed pick, or a pick with a point at one end and a chisel edge at the other end.

Needle penetrometer
See **Proctor needle.**

Negative adsorption
When uptake of cations occurs from an electrolyte but the clay has little or no sorption capacity for the anions and their concentration in the intermicellar solution rises.

Negative mantle friction
See **Negative skin friction.**

Negative skin friction
That which causes the bearing capacity of a pile driven into a soil to decrease during the course of time.

Net bearing pressure
See **Net loading intensity**.

Net loading intensity
The additional intensity of vertical loading at the base of a foundation due to the mass of the new structure, including earthworks (if any). Usually it is the difference between the gross loading intensity before building operations are commenced and the gross loading intensity after the structure is complete and fully loaded.

Neutral pressure (u) (or **neutral stress**)
The hydrostatic pressure in the pore water of the soil. Nowadays always called pore water pressure.

Neutron moisture meter
See **Moisture probe.**

Newlyn datum
The Ordnance Survey levelling datum in Cornwall.

Newmark chart
An influence chart after N M Newmark for computing vertical, normal stress at depth caused by a uniformly distributed load at the elevation of the base of the footing of a foundation.

Newton (N)
A newton is the force which acting on a body of mass 1 kilogram will accelerate it 1 m/s^2.

Newtonian liquid
A perfectly viscous material.

Newtonian model
A simple rheological model to represent a perfect viscous response of a material.
See also **Rheological model.**

NGI
Norwegian Geotechnical Institute.

NOD
Newlyn Ordnance Datum.

Non-cohesive soils
These include the coarser and largely siliceous and unaltered products of rock weathering. They possess no plasticity and tend to lack cohesion especially when in the dry state.

Non-displacement pile
See **Bored pile.**

Non-flowing artesian
See under **Artesian water.**

Non-particulate grout
See under **Grout.**

Non-plastic (NP)
A soil with a plasticity index of zero or one on which the plastic limit cannot be determined.
See also **Plasticity index.**

Non-saturated flow
Laminar flow of liquid through a soil, the transfer mechanism of which depends on the degree of saturation.
See also **Saturated flow.**

Normally consolidated soil
Soil which has been subjected only to an increase of pressure.

Normally loaded
See **Normally consolidated soil.**

Normal stress (σ_n)
The stress in a soil normal (perpendicular) to the shear stress.

North Dakota cone
A shaft with a sharp cone of angle 15½° at one end. Used as an *in situ* soil penetration test to determine the pavement thickness required on the sub-soil.

NP
Symbol for non-plastic.

Nuclear densometer
An instrument containing a radio-active source for determining, *in situ*, the bulk density of surface layers of soil. Two basic types are the direct transmission and back scatter.
In the direct transmission apparatus a radio-active source, producing the gamma radiation, near the end of a thin probe, is inserted into the soil to be tested. The intensity of radiation emerging at the surface is measured by a detector placed on the surface of the soil and connected to pulse counting equipment. The higher the bulk density of the soil the greater is the absorption of gamma radiation.
In the back scatter apparatus both the radio-active source and radiation detector are contained in a unit which is placed on the surface of the soil. Some of the gamma radiation entering the soil below the source is scattered back and detected by the Geiger tube. The intensity of radiation detected decreases with an increase in the bulk density of the soil.
See also **Density probe** and **Geophysical exploration**.

Null indicator
This is basically a U-tube of small-bore glass containing mercury in the bottom half of each stem. The stems are connected to two different pressure sources by rigid tubes and when the mercury is made to be level across the two stems of the U-tube the pressures are balanced.
See also **Perspex null indicator**.

N-value
Unit of penetration resistance. Also stability number for design of slopes.

O

OBM
Bench mark.

Observation survey
See **Preliminary survey**.

Odour
The smell of decaying vegetation given off by recently excavated soil thus showing it to contain a high proportion of organic material.

Oedometer
A consolidation test apparatus.

Ogee joint
A type of spigot-and-socket joint in a drain pipe.

Oil dash pot
See **Pressure system**.

Old rail grillage
See **Rail crib**.

One-dimensional consolidation
Settlement where all the strain in the soil is in the vertical direction only.
See also **Consolidation** and **Three dimensional consolidation**.

One-way drainage
See under **Drainage path**.

Open caisson
A caisson open both at the top and at the bottom.
See also **Caisson**.

Open capillary fringe
See under **Capillary water**.

Open-ended pipe grouting
This is the simplest form of grouting using an open-ended injection pipe. Whilst being installed the pipe is sealed at its lower end with a plug which is expendable once grouting is started.
See also **Grouting**.

Open sheeting
Used in excavation in which the sides are reasonably firm and not likely to crumble. Generally consists of vertical poling boards spaced at intervals and supported by walings and struts, or horizontal sheeting openly spaced and held in position by soldiers and struts.

Open standpipe piezometer
See **Standpipe piezometer**.

Optimum density
See **Maximum dry density**.

Optimum moisture content
The moisture content of a soil at which the precise amount of compaction produces the highest dry density.
See also **Maximum dry density**.

Orange peel bucket
A grab shaped like a half orange skin cut into segments.
It is used for excavating gravel and sand.

Ordinary well
See under **Well**.

Ordnance datum
See **Newlyn datum**.

Organic clay
See under **Organic silt**.

Organic fine soil
Fine soil containing a substantial amount of vegetable organic matter.

Organic silt
Some lakes, reservoirs, bays, estuaries and harbours are today receiving sediment carried in by rivers. The depositing of this material is said to result in

silting-up. This use of the term silt, therefore, has no reference to the particle size of the material being deposited, which is, in fact, chiefly a soft mud of material of the clay grade, though silt and even sand grade material may be present. These deposits are often accompanied by a considerable amount of organic matter, which makes itself evident by odour when the deposit is disturbed. Material so formed, which includes most recent alluvial muds, are referred to as organic silts or organic clays. The organic matter present renders these soils more plastic and smoother to the touch than deposits of similar texture which contain no organic matter.
See also **Silt**.

Organic soil
A soil whose physical properties are strongly influenced by the presence of organic matter.

Oriented core barrel
A borehole surveying instrument which takes a core and at the same time records the bearing and the slope of the borehole.

Oslo point
A steel bar, 75 to 100 mm in diameter, protruding from the bottom of a driven pile. The lower end of the round bar is hollow-ground and hardened to prevent the point of the pile skidding over a hard rock strata.

Osmosis
The diffusion of a solvent or of a dilute liquid through a skin (permeable in only one direction) into the more concentrated solution. It is analogous to the movement of water in soil during electro-osmosis.

Osmotic pressure
The pressure is osmosis exerted by a solvent when its entry through the skin into the more concentrated solution is prevented.

Osmotic suction
See **Solute suction**.

Osterberg sampler
A piston soil sampler developed by Osterberg in 1952. Basically it consists of a hollow drill rod and lower larger diameter sample tube. Inside the sample tube is a further thin-walled piston sampler which can be pushed forward and retrieved by hydraulic pressure passing through the hollow drill rods.

Outcrop
An exposure of a stratum at the earth's surface.

Outflow
See under **Seepage**.

Outward seepage
See under **Seepage**.

Outwash fan
See **Alluvial fan**.

Oven dry soil
Soil dried in an oven at 105°C for 24 hours.

Overburden pressure
The vertical pressure at a point in a soil mass due to the mass of material above it.
See also **Effective pressure** and **Total pressure**.

Overcompaction
See **Optimum moisture content**.

Overconsolidated soil
A soil that has been consolidated under greater pressure than that provided by its present overburden.

Overconsolidation ratio
The ratio of the preconsolidation pressure divided by the present overburden pressure.

Overdriving
Continuing to drive a pile into the ground after its base has reached a satisfactory hard bearing stratum.

Overfill
Extra fill material placed over the permanent fill material of, say, an embankment, to accelerate consolidation. Once measurements indicate that consolidation is nearly complete the overfill is removed.

Overflow settlement cell
See **Hydraulic overflow settlement cell**.

Overload fill
See **Overfill**.

Overload ratio
The ratio by which the yield value of a clay, as found in a slow undrained shear test, is multiplied in order to obtain a permissible value for use in design.

Overshelling
The formation of a cavity, by the collapse of the wall of a borehole, when boring through non-cohesive soil and due to the borehole lining tubes not being advanced down the borehole deep enough.

Oxidation
The process whereby oxygen is added to the soil from the action of air aided by the presence of moisture.
Iron-containing minerals are commonly subjected to oxidation. The products of oxidation are split off and some are washed out of the soil.

P

Packer test
An *in situ* permeability test carried out at intervals in an advancing borehole. The test is carried out by sealing off a length of uncased borehole with either hydraulic or mechanical packers, and injecting water under pressure into the isolated test section. The rate of flow of water over the test length is measured under a range of constant pressures.

Packing
A pad of resilient material contained between the helmet and the top of a reinforced concrete pile to minimise damage to the head during driving.

Packing geometry
The way in which the particles of a soil are packed together.

Pad foundation
An isolated foundation to spread a concentrated load, such as from a column.

Page
A small wooden wedge used in timbering trenches.

Paired consolidation test
See **Swelling test.**

Palm pile
A pile developed by the Belgian, Jean Rooseu, in 1961. It consists of a cylindrical shell driven on top of a precast concrete tip. Just above the tip, tucked into the shell, are four steel pivot plates. These plates lie within the shell during driving, but when a bearing stratum is reached, the cylinder is slightly raised, the plates (or palms) are mechanically thrust out into the bearing stratum, and the shell is then filled with concrete. The effect of the palms is to double the bearing value of the pile.

Pan mixer
See **Mixer.**

Paraffin wax
A white crystalline substance obtained from shale and when mixed with small quantities of beeswax, carnaubawax and ceresine is used to seal soil samples that are to be stored.
The paraffin wax is applied at a temperature as near as possible to its congealing temperature.

Partially saturated density (γ)
The density of a soil when the voids are only partially saturated.
It is given by:

$$\frac{(Gs + Se)\gamma_{\omega}}{1 + e}$$

where Gs = specific gravity of the soil particles,
S = degree of saturation,
e = voids ratio,
and γ_{ω} = density of water.
See also **Density.**

Partially saturated flow
See **Non-saturated flow.**

Partial saturation
When the voids of a soil are partly filled with air (or gas) but the water in the voids is continuous.

Particle
The individual minutest grain into which a soil can be divided.

Particle shape
The particle shape of sand-grains or pebbles may be described as angular, rounded, subangular, subrounded or irregular.

Particle size distribution
A particle-size distribution expresses the size distribution of the particles comprising a soil.
See also **Grading curve**.

Particle size distribution chart
A logarithmic-based chart on which to conveniently plot the particle-size distribution curve of a soil sample.

Particle specific gravity
See **Specific gravity**.

Particulate grout
See under **Grout**.

Pass
The number of passes is the number of times that each point on the surface of a layer of material being compacted has been traversed by the item of compaction plant.

Passive earth pressure (p_p) (or **passive resistance**)
The resistance of a soil face to deformation by a horizontal force (usually due to active earth pressure). The passive earth force in sand is equivalent to that of a fluid weighing three to four times as much as the sand. It is the reciprocal of (p_a), the active earth pressure in Rankine's theory.
See also **Coefficient of passive earth passive**.

Paulding clay
A deep water deposited clay derived from drift, consisting mainly of shale.

Pavement
The part of a road or airfield runway which distributes the wheel loads over an area so that the bearing capacity of the underlying soil (subgrade) is not exceeded. It usually consists of two or more layers of material: a top layer or wearing surface which is durable and water-proof, and a base material, the base material sometimes being further split into two layers—a base course and a sub-base course.
Pavements can be either of the flexible or rigid type depending mainly on the local economic considerations. Flexible pavements can have lean concrete bases, cement bound granular bases, tar or bitumen bound macadam, all overlain with bituminous surfaces. Rigid pavements consist of reinforced concrete base and surface.

Pavement logger
See **Road logger**.

Pavement settlement point
An apparatus to enable measurements of settlement and heave of a pavement

to be monitored by normal surveying practice.
It consists of a steel tube 32 mm in diameter, internally threaded, with a fixed base plate and removable cover plate. This small chamber is securely fixed just below the surface of a pavement. When it is required to monitor the pavement level, the cover plate is removed and a surveying reference point is screwed into the recess.

Paving
A simple road surfacing constructed directly on natural ground using precast concrete slabs, asphalt or coated macadam.
See also **Pavement**.

Paving flags
Hydraulically formed paving slabs consisting of a mixture of 1:3 cement:granite chippings and dust. They are usually made in standard sizes to BS 386 and are suitable for laying on a thin bed of sand directly on the soil to form a walkway.

Pay gravel
A large quantity of gravel which can be excavated and sold at a profit.

Pea gravel
Gravel having the same sized particles as peas.

Peak strength
See under **Residual strength**.

Pea shingle
See **Pea gravel**.

Peat
An accumulation of fibrous or spongy-textured vegetable matter formed by the decay of plants more or less *in situ*. True peat is combustible and composed almost entirely of organic matter, though inorganic material may be present in varying amounts and, when excessive, may result in organic clays or silts. Peats are classified as high or moor peats or as low or fen peats. The latter, which may form on any alluvial tracts, are often interbedded with and sometimes buried beneath layers of alluvial mud. In the agricultural sense a peaty soil is a topsoil with a high humus content. From the engineering standpoint, a peat soil may be a few metres thick and is characterised by a high degree of compressibility.

Pebble gravel
See **Gravel**.

Pebbles
A constituent part of gravels.

Pedological soil
See under **Soil**.

Peg
A short pointed wooden stave driven into the ground to mark a line or level.

Pendulate water
Water trapped in the voids of a soil which is above the ground-water level.

Pendulum apparatus
See **Hanging** and **Inverted pendulum**.

Penetration

Penetration needle
See **Proctor needle.**

Penetration rate
See **Rate of penetration**.

Penetration tests
Tests of the soil *in situ* which give an indication of its load-bearing capacity and density. They can be divided into static and dynamic tests. Evidence from penetration tests should always be interpreted with the help of information from boreholes.
See also **Split spoon sampler**.

Penetrometer
A cone-shaped instrument which is pushed into the ground to the required level and then forced in under a measured pressure.

Percentage air voids (V_a)
The volume of air voids in the soil expressed as a percentage of the total volume of the soil.

Perched water table
This occurs if a permeable water-bearing soil layer overlies an impermeable stratum which is itself above the normal ground-water level.

Percolation
The movement of gas, oil or water through the pore spaces of the ground.

Percussion boring
A method of boring a hole vertically through soil by the use of a free fall mass being dropped onto the top of a shell auger.
The apparatus consists of a portable tripod which sits over the position of the borehole, and a diesel-driven winch which is used to repeatedly raise the anullus 'free fall mass' which runs down the drill rods fixed to the auger.

Percussion drilling
Using a heavy drilling bit and alternately raising and dropping it so that it grinds the underlying soil to the consistency of sand, so forming a borehole through hard soil or rock. During this process the borehole, if possible, is kept dry except for a little water that forms a slurry with the disturbed soil. The slurry is removed by bailer.

Pereletok
Description of a soil (or rock) that has been frozen for two seasons.
See also **Frozen soil.**

Perenially frozen soil
Soil that has been frozen from at least three years to several decades.
See also **Frozen soil.**

Perforated drain
Perforated drain pipes placed in a trench and surrounded with granular material satisfying the requirements of a filter. The ideal width of the openings in the drain pipe should be equal to the 60% (D_{60}) size of the surrounding material.

Period of rest
See **Rest period.**

Periscope
See **Borehole periscope.**

Permafrost
Soil which has been frozen for centuries such as is found in Alaska, Northern Canada and Siberia.
See also **Frozen soil.**

Permanent reference studs
See under **Reference studs.**

Permanent set
The permanent distortion of a structural member due to over-stressing.

Permeability
The rate at which water flows through a soil under unit hydraulic gradient.

Permeability coefficient
See **Coefficient of permeability.**

Permeable blanket
See **Drainage blanket.**

Permeameter
A laboratory instrument for measuring the coefficient of permeability of a soil sample. The constant head permeameter is used for permeable materials such as sand or gravel, the falling head permeameter is used for impermeable materials such as clay or fine silt.

Perspex null indicator
A null indicator using a water mercury interface in a U-type configuration. It consists of a small block of Perspex which has been drilled to form a V-tube, together with a third drilled tube at right angles to the top of one of the tubes forming the V. This third tube acts as a trap for the contained mercury in the V-tube when the indicator block is pivoted about a horizontal axis for de-airing and flushing water through the triaxial cell test apparatus.
See also **Null indicator.**

Pervious
Permeable.
See also **Impervious.**

Pervious blanket
See **Drainage blanket.**

Petrification degree
The ratio of the shrinkage limit of a soil to its absorption limit. The higher the value the better the prospect of self-stabilisation.

PFA
Pulverised fuel ash.
See **Fly-ash.**

pF scale
A scale developed by R K Scofield for the measurement of soil suction. Because of the wide range, it is convenient to measure suction on a logarithmic scale. The pF index is therefore defined as:—

$$\log_{10} \text{(total suction, measured in cm of water)}.$$

Phase
See **Soil phases**.

Phenoplastic
A thermosetting plastic based on phenols and used in soil grouting.

pH meter
An electrically operated instrument with a dial scale for the direct measurement of soil pH.
See also **Barium sulphate test**.

Photoelastic load indicator
This consists of a cylinder of optical glass placed in a hole in a steel cell body, load being applied through end plates. These plates are usually domed and are arranged to accommodate eccentricity of loading. Spring washers retain the loading plates in position under zero load.
The principal of operation is that when the cylinder of optical glass is strained under load, photoelastic interference fringe patterns are visible in the glass cylinder when it is illuminated with polarised light. The fringe pattern observed is a direct measure of the load applied to the load cell.

Photogrammetry
A standard surveying technique which can be used to study the movement of structures.
The method requires the use of a photo-theodolite, or some other precise camera, which is used to take a series of successive photographs from a fixed station along a fixed camera axis. Where the movements are in a plane parallel to the image plane of the camera, one camera station is used. To measure three-dimensional movement it is necessary to photograph from two camera stations with the lines of sight convergent.

Photographic inclinometer
An instrument for monitoring soil movement. It consists of a steel cable attached to the top of a brass container, and the cable and container are marked at 1 m intervals. Inside the container are an electrically-operated timed camera, a pendulum, a light source and a compass. The container is lowered down a borehole 1 m at a time with a time interval of one minute between each increment of depth. The camera takes a photograph of the pendulum and compass face, thus giving the inclination of the borehole and direction.

Photo-theodolite
See **Photogrammetry**.

Phreatic surface
Ground water level.

Phreatic water
See under **Saturation zone**.

pH scale
A scale developed by Sorensen for the measurement of acidity in a liquid.

PI
Symbol for plasticity index.

Pick
A digging tool with two long, sharp points, used for breaking loose rock or digging stiff clay.

Pick axe
See **Navvy pick**.

Pier
(1) An underground structural member, having a maximum depth to width ratio of 4:1, that serves the same purpose as a footing.
(2) A bridge support, usually of concrete or masonry, rising above ground level or river water level. Under this definition a pier must be provided with its own foundation.

Pier shaft
That part of a pier above its foundation.

Piezometer
See **Piezometric tube**.

Piezometer tips
These are used to record the piezometric level in saturated or partly saturated soil, or other material. They are connected to an access tube and normally placed in a sand column at the bottom of a borehole, or other suitable location, and then sealed from the surface. The piezometric head is determined by recording the water level in the access tube with a dipmeter.
Alternatively the access tube can be capped and connected by two piezometer tubes to a remote reading location.
The various types at present available are listed below:—

Acoustic
This comprises a high or low porous entry ceramic element integral with a diaphragm vibrating wire transducer in a stainless steel cylindrical casing. Overall length is 240 mm, with a diameter of 38 mm.

Bishop
This piezometer tip is for use in partially or fully saturated soil or other material, and is suitable for positive and negative pressure. It was designed by Professor A W Bishop. It comprises a high air entry ceramic element cemented to single brass end cap fittings to take two piezometric tubes. The ceramic element is 100 mm long, tapering from 50 mm to 38 mm diameter at the bottom.

Bull-nosed
This comprises a high air entry ceramic element cemented to single brass end cap fittings to take two piezometric tubes. The element is 100 mm long, 50 mm in diameter and has a domed nose. This tip is suitable for use in fully or partially saturated soil and for positive and negative pressures.

Piezometer

Cambridge drive-in
This consists of a cylindrical low air entry porous plastic element protected during driving by a conical-ended steel guard which is driven off by a mandrel after the tip has reached the correct depth. It is suitable for driving in soft saturated soil. The tip is 1.08 m in length and 27 mm in diameter.

Casagrande ceramic
This consists of a cylindrical low air entry porous ceramic element with rigid plastic end fittings held by a central perforated plastic tube, fitted with a plastic connector to fit a PVC standpipe. The tip is 250 mm long by 50 mm diameter and is suitable for installation in boreholes.

Casagrande drive-in
This comprises a cylindrical low air entry porous plastic element protected by a perforated galvanised mild steel pipe with cadmium plated mild steel end cone. The tip is 300 mm long by 43 mm overall diameter. The tip is driven into soft saturated soils by means of its mild steel standpipe.

Casagrande porous plastic
This comprises a cylindrical low air entry porous plastic element protected by a perforated rigid plastic tube with rigid plastic end fittings to fit onto a plastic standpipe. The tip is 300 mm long by 32 mm overall diameter. It is suitable for installation in boreholes.

Cylindrical
This comprises a coarse low air entry ceramic element cemented between plastic end caps which have fittings for two piezometer tubes. The element is 130 mm long and 50 mm diameter and is suitable for use in fully saturated soil.

Disc
This comprises a high air entry ceramic element bonded into a plastic body with brass fittings for two piezometer tubes. The element is 100 mm in diameter and 6 mm thick and is suitable for positive and negative pressures in fully or partially saturated soil.
See also **Boundary piezometer**.

Hydraulic/pneumatic
This comprises a brass transducer body incorporating a stainless steel diaphragm, for medium and high pressures, and fitted with two brass connectors to take piezometer tubes. The transducer is fitted with a porous element clamped by a brass nose cone threaded onto a central spindle. The tip is 270 mm long and 38 mm in diameter.

Mini
This comprises a coarse low air entry porous plastic element supported and clamped between brass end caps, with two fittings for twin piezometer tubes. It is only 63 mm long and 19 mm in diameter.

Push-in
This comprises a ceramic element retained between plastic end caps, one cylindrical and one conical, with a central brass reinforcing rod protruding to locate placing adaptor. Depending on the type of element used the tip can be used in saturated or partially saturated soil or in soil containing gas. It is 250 mm long and 40 mm in diameter.

Pneumatic
This tip is similar to the Hydraulic/pneumatic tip except that it incorporates a viton diaphragm for medium and low pressures.

Piezometric head
The vertical head of water in a piezometric tube.

Piezometric level
The level of water in a piezometric tube.

Piezometric tube
A tube, having at its lower end a special piezometric tip, which is inserted into soil to obtain in the tube a pressure equal to the pore pressure existing in the soil.

Pile
A large 'stake' driven deeply or formed deeply in the ground for transmitting the mass of a structure to the soil by the resistance developed under its base and by friction along its surface. For the different types of pile available see under **Foundation**.
For other topics associated with piling see under the following headings:—

Adhesion factor
Bootstrap test
Converse-Labarre formula
Deflected pile
Dolly
Driving band
Driving cap
Drop hammer
Electrochemical hardening
Engineering News formula
Heaved pile
Helmet
Hiley formula
Kentledge
Lateral pile load test
Long dolly
Maintained load test
Meyerhof's formula
Negative skin friction
Overdriving
Packing
Pile cage
Predrilling
Rig
Set
Skin friction

Pile adhesion
See under **Adhesion factor**.

Pile anchor
See **Anchor block** and **Tension pile**.

Pile bent abutment
See **Spill through abutment**.

Pile cage
Prefabricated rigid steel reinforcement consisting of main longitudinal bars and secondary hoops, which is lowered centrally into a bored hole before concreting to form a bored pile.

Pile cap
A concrete block cast on the head of a pile or a group of piles to transmit the load from the structure to the pile or group of piles.

Pile capacity
See **Ultimate pile load**.

Pile cluster
See **Pile group**.

Pile cushion
See **Dolly**.

Piled footing
A reinforced concrete footing supported on piles. It thus also acts as a cap for a group of piles. The piles normally project 75 mm into the footing.

Piled foundation
A foundation carried on piles to suitable ground beneath the surface.

Pile drain
See **Vertical sand drain**.

Pile drawer
See **Pile extractor**.

Pile driver
A frame to locate a pile and to guide it during the initial stages of penetration, to support and guide the hammer, and to carry the power-operated winch.

Pile extractor
A special piece of equipment attached to, or gripping, the top of a driven pile to remove and salvage it from the soil. The extractor can be operated by (1) direct pull (2) hydraulic jack, (3) inverted double-acting hammer, (4) ultrasonic means or (5) vibrating action.

Pile force gauge
A vibrating wire load gauge specifically designed by the Swedish Geotechnical Institute for measuring both tensile and compressive forces in piles. It is also capable of withstanding the forces due to pile driving.

Pile freeze
See **Freeze**.

Pile friction
See under **Skin friction**.

Pile group
Several driven or bored piles placed close together to take a heavier load than a single pile could carry. A pile group, whether of steel, concrete or timber piles, is generally capped by a reinforced concrete pile cap.

Pile heave
See **Heaved pile**.

Pile hoop
See **Driving band**.

Pile jetting
See **Pile water jet**.

Pile load test
A load test on one or more piles driven at the start of a large piled project. The piles are load tested to check their suitability for the project and must therefore

be of the same type and driven by the same equipment as anticipated for the project.

Pile overdrive
See **Overdriving**.

Pile pulling test
A test to find the pull-rise curve of a friction pile to supplement the load-settlement curve. As both curves have the same characteristics for a particular friction pile, the difference between the failure load determined from the load-settlement curve and the pull-rise curve gives an indication of the pile end bearing resistance.

Pile redriving test
A test to ascertain whether end bearing piles already driven to a satisfactory bearing stratum need redriving.
When further nearby piles are driven, earlier driven piles may be lifted off the bearing stratum due to the upward movement of the soil displaced by the additional piles. To determine if redriving is required, it is necessary to establish the level of the top of each pile after it is driven and to recheck the level after all piles have been driven.

Pile relaxation
See **Relaxation**.

Pile resistance
See **Set**.

Pile ring
See **Driving band**.

Pile shoes
Small pointed blocks of cast-iron or steel, fixed to the bottom ends of timber and concrete piles to protect them during driving through different types of soil strata.

Pile spudding
See **Spudding**.

Pile water jet
To assist pile driving into compact sand a water jet is provided ahead of the pile toe. The water is led from a suitable pump through a hose to a 40 mm diameter pipe, embedded into the concrete pile and terminating in a 10 mm diameter nozzle in the pile toe. With timber piles the water pipe is fixed to the side of the pile.

Pillar drain
See **Buttress drain**.

Pilot pile
A heavy duty steel pile used to drive through obstructions such as boulders, riprap and timber etc before driving lighter piles made of concrete or timber.

Pilot trench
See **Guide walls**.

Pinchers
A pair of poling boards strutted apart to support a trench wall in good soil.

Pipe pile
A pile consisting of a length of steel pipe, driven either open-ended or with a shoe.

Pipette analysis
See **Fine analysis**.

Piping
Subsurface erosion or boiling caused when the pressure due to the flow of water is equal to, or greater than, the effective pressure between the soil particles.

Piston sampler
An apparatus to obtain undisturbed samples of soft or slightly cohesive soils. It consists of a thin-walled tube fitted with a piston that closes the end of the sampling tube until it is lowered into a borehole. The sampling tube is then pushed past the piston into the soil. During withdrawal of the sampler the piston prevents water pressure from acting on the sample.
See also **Pocket piston sampler**.

Pit
In the context of soil mechanics a pit is an excavation ranging from that required for a trial hole to an excavation to receive the basement and foundations of a building.
A pit up to 1.5 metres deep is considered as shallow, between 1.5 and 4.5 metres deep as medium, and over 4.5 metres as a deep pit.

Pitcher sampler
A rotary barrel sampler in which a sampling tube is suspended inside the outer cutter barrel while it is being lowered to the bottom of the borehole. The sampling tube is then forced into the soft soil ahead of the cutter barrel by a spring.

Pitching
(1) Bricks, concrete, concrete blocks and stones etc, grouted where necessary, placed on the surface of a soil slope to protect it from erosion.
(2) The process of raising runners or piles into vertical position ready for driving into the ground.

PL
Symbol for plastic limit.

Plain sinker bar
See under **Sinker bar**.

Plane of rupture
The plane along which retained soil is imagined to fail when a wall is being designed to retain it using the wedge theory. The angle of elevation of the plane determines the mass of the soil retained. Also applies to plane of failure in any shearing situation.

Planer
A machine of the grader type fitted with a series of blades for distributing, smoothing and grading loose surfaces.

Plant
Engineering plant comprises any power-operated machine or tool used primarily for engineering construction work.

Plant mix
Soil stabilisation by carrying the soil to a stationary mixer, returning it to the site and respreading it.

Plaster pad
See under **Telltale**.

Plastic failure
A compression failure with well expressed lateral bulging of a soil sample in an unconfined or triaxial compression test.
See also **Brittle** and **Intermediate failure**.

Plastic flow
See **Creep**.

Plastic frozen soil
See under **Frozen soil**.

Plasticity
That property which enables a soil (or other material) to be deformed continuously and permanently without rupture.

Plasticity assessment
See **Rapid assessment of plasticity**.

Plasticity chart
A chart produced by Arthur Casagrande which, if the liquid limit and plasticity index of a soil are known, gives an indication of the soil properties to be expected.
In the plasticity chart the ordinates represent the plasticity index of a soil, and the abscissas the corresponding liquid limit of the soil.
The chart is horizontally divided into three sections by two vertical lines C & B at liquid limit values of 30 and 50 respectively and is further divided into six regions by a sloping A line across the chart.
The group to which a given soil sample belongs is determined by the name of the chart region that contains the point representing the values of plasticity index (PI) and liquid limit (LL) for the soil. All points representing inorganic clays lie above the A line, and all points for inorganic silts lie below the A line.

Plasticity index (PI)
The range of water content over which a cohesive soil behaves plastically; ie, the range lying between the liquid and the plastic limits is defined as the plasticity index. Thus:

$$\text{plasticity index} = \text{liquid limit} - \text{plastic limit}.$$

See also **Atterburg limits**.

Plastic limit (PL)
The water content at the lower limit of the plastic state of a clay. It is the minimum water content at which a soil can be rolled into a thread ⅛″ (3 mm) in diameter without crumbling.
See also **Atterburg limits.**

Plastic membrane
See under **Membrane.**

Plastic permeameter
A permeameter constructed entirely of transparent plastic to enable observations to be made of soil behaviour and water flow during a demonstrated permeability test.

Plastic piezometer tip
See under **Piezometer tip.**

Plastic range
See **Plasticity index.**

Plastic viscosity
The shearing stress in excess of the yield value that will induce a unit rate of shear. In soil mechanics plastic viscosity is a measure of that part of slurry flow resistance caused by mechanical friction.

Plastic water stop
See **Surface waterbar.**

Plastometer
A device similar to an oedometer but equipped with peripheral electrical strain gauges.
The measurement of strains in the oedometer metal ring enables the determination of average radial stresses on the periphery of the soil sample.

Plate-bearing test
A method of estimating the bearing capacity of a soil by digging a pit down to the proposed foundation level, placing a stiff steel plate about 300 mm square on the foundation, and loading it until it fails by sinking rapidly.

Plugging
Sealing off an abandoned borehole.

Plumbness
Vertical alignment—especially of a drilled hole or placing a pile.

Plunger head
The circular plate in contact with a soil sample being tested in a compression testing machine.

Plunger rod
The rod connecting a proving ring to a plunger head in a compression testing machine.

Pneumatic caisson
See **Compressed-air caisson.**

Pneumatic extensometer
See under **Potentiometric extensometer**.

Pneumatic/hydraulic pressure cell
A cell consisting of a circular or rectangular flat jack formed from two sheets of steel welded around the periphery. The narrow gap between the plates is filled with oil. The cell is connected to a pneumatic or hydraulic transducer by a short length of steel tubing, forming a closed hydraulic system. Both cell and transducer are embedded in the structure to be monitored. Twin nylon tubes connect the transducer to a terminal read-out unit. To take readings air, nitrogen or hydraulic oil pressure is supplied from the remote read-out unit to one side of a flexible diaphragm valve incorporated in the transducer. When the supply pressure is sufficient to balance the cell pressure on the reverse side of the diaphragm, the valve opens, allowing flow along the return line to a detector in the read-out unit.

Pneumatic piezometer
The piezometer tip comprises a porous element integral with a diaphragm transducer, installed either in a borehole or by pushing to shallow depths in soft soil. Twin piezometer tubes connect the transducer to a remote read-out terminal. To take pressure readings air, nitrogen or hydraulic fluid pressure is supplied from the terminal to one side of the flexible diaphragm valve incorporated in the transducer. When this applied pressure is sufficient to balance the groundwater pressure acting on the reverse side of the diaphragm, the valve opens, allowing flow along the return line to the read-out terminal.

Pneumatic piezometer tip
See under **Piezometer tips**.

Pneumatic settlement cell
A cell for the measurement and control of vertical movement.
It consists of a pneumatic transducer with an integral mercury chamber. Triple nylon tubes extend from the cell to a remote read-out unit located in stable ground. One of the tubes is filled with mercury and is connected at one end to a datum sight tube and at the other end to the mercury chamber. Vertical movement of the cell relative to the read-out location results in a change in mercury pressure within the chamber. To measure the pressure change, and hence the amount of settlement or heave, pneumatic pressure is supplied from the read-out unit to one side of a flexible diaphragm valve incorporated in the cell transducer. When the supply pressure is sufficient to balance the mercury pressure on the reverse side of the diaphragm, the valve opens allowing flow along the return tube to a detector in the read-out unit.

Pneumatic tyred roller
A rubber-tyred self-propelling heavy roller having four or more tyres abreast, back and front, used to compact slightly cohesive granular soil at or near its optimum moisture content.

Pocket dipmeter
See **Dipmeter**.

Pocket penetrometer
A small instrument for the accurate classification of cohesive soils in terms of consistency.

Pocket pH meter
See **pH meter** and **pH scale.**

Pocket piston sampler
A small piston sampler light enough to carry in a briefcase. It is designed for taking soil samples 25 mm in diameter and 50 mm long during examination of soil strata. It can also be used for taking small control samples or test specimens from large undisturbed samples.
See also **Piston sampler**.

Pocket shear meter
See **Torvane.**

Podzol
A relatively acid surface soil from which much soluble material (iron and aluminium oxides) has been leached into underlying soils. It has more silica than iron and aluminium oxides together, and may have up to seven times as much.

Point bearing pile
See **Bearing pile.**

Point load
A load concentrated at one point.

Poise
A unit of measurement equal to 1 dyne/s/cm^2.

Poiseuille's law
The amount of discharge is directly proportional to the square of the tube radius.
This refers to experiments carried out on the flow of soil moisture which may be visualised as a flow in a system of small tubes corresponding to soil pores.

Poissons ratio
If a rod of elastic material is stretched (or compressed) lengthwise with sufficient pull (or push) it will be elongated (or decreased in length). The unit of change in length per unit of original length is strain. At the same time the lateral dimension will change.

$$\text{The ratio } \frac{\text{lateral strain}}{\text{longitudinal strain}} = \text{Poissons ratio.}$$

Polar forces
The forces set up when two molecules, where the positive and negative charges are not concentric, approach each other.

Polder
Low-lying land reclaimed from the sea in Holland.

Poling back
The operation of excavating behind 'timber' supports already in position and 'timbering' the new face.

Poling board
A flat member in contact with the ground and supporting the face or sides of an excavation, usually 1 m to 1.5 m long.

Poling frame
A frame in which the walings support the poling boards at their mid-points.

Polyester
A thermosetting plastic used in soil grouting.

Polymer stabilisation
Mixing polymers with catalysts and retarders and injecting the solutions into the soil to be stabilised. After a period of time, depending on the amount of retarder added, the solution forms a gel which is virtually impervious.

Polysaccharides
Carbohydrates which are found naturally as alginates and used in soil grouting.

Poorboy
The simplest type of core sampler, so called because it is made of any available tube with suitable diameter and wall thickness when normal core samplers are not readily available.

Poorly graded
A sample of soil is said to be poorly graded when it has a deficiency in some intermediate-sized particles, or an excess of some other sized particles.
See also **Coefficient of uniformity**.

Pore air pressure
The pressure of the air in the voids of a partially saturated soil. This pressure may not be the same as the pore water pressure in the same soil due to the surface tension at the air water interfaces.

Pore pressure dissipation
See **Dissipation test**.

Pore pressure excess
See **Excess pore pressure**.

Pore pressure parameters
The parameters A and B are best understood by considering the case of a soil sample in a triaxial compression test. An increase in the principal stresses σ_1, σ_2 and σ_3 will result in a decrease in volume and a consequent increase in pore pressure of the sample under test, the decrease in volume of the soil sample being almost entirely due to the decrease in volume of the voids.
The changes in pore pressure are expressed in terms of empirical parameters A and B such that

$$\Delta u = B\,[\Delta\sigma_3 + A\,(\Delta\sigma_1 - \Delta\sigma_3)]$$

where Δu represents an increase in pore water pressure
and $\Delta\sigma_1$ and $\Delta\sigma_3$ represent increases in the principal stresses.

For a fully saturated soil the A parameter depends mainly on whether the soil is normally consolidated or overconsolidated and on the proportion of the failure stress applied. The B parameter is unity.
For a partially saturated soil where there is some air in the voids, the value of the B parameter is less than unity and varies with the stress range.

Pore pressue ratio
The ratio, at any given point in a soil mass, of the pore water pressure to the mass of the material acting on unit area above it.

Pore pressure transducer
See **Pressure transducer**.

Pores
Small cavities in soil, particularly granular soil like sand, ie the voids in a soil.

Pore-water pressure (u)
The pressure of water in the voids of a soil.
See also **Excess pore pressure** and **Pore air pressure**.

Pore-water pressure cell
Sensitive non-ferrous instrument for the *in situ* measuring of pore water pressure.

Porosity (n)
The ratio of the volume of voids to the total volume of a soil sample. In sands it is from 30% to 50% but in clays may be above 90%. If the voids ratio is e, the porosity n = e/1 + e.
See also **Effective porosity**.

Porous plastic piezometer tip
See under **Piezometer tips**.

Portable clinometer
An instrument to measure superficial movement such as tension cracks and fissures.
This comprises a measuring beam with a hardened V-strip along its base. The V-strip locates on one of a pair of reference ball studs set into the structure or ground. A conical recessed plunger at one end of the beam locates on the second reference ball. Screw adjustment at the plunger position brings the beam into a horizontal plane; bubble levels mounted on the beam indicate when adjustment is complete. A depth gauge is then inserted through the plunger to register the difference in elevation between the two balls.
See also **Portable extensometer, Portable rod micrometer** and **Reference studs**.

Portable dipmeter
See **Dipmeter**.

Portable extensometer
An instrument to measure superficial movement in structures and foundations. It comprises a tube extensometer. Hardened conical end pieces locate on ball reference studs set into the structure or foundation and an integral dial depth gauge used to obtain the relative movement between the studs.
See also **Portable clinometer**, **Portable rod micrometer** and **Reference studs**.

Portable penetrometer
See **Pocket penetrometer**.

Portable rod micrometer
An instrument to measure superficial movement in structures and foundations. It comprises a micrometer head with extension rods. Hardened end pieces, one conical and one flat, locate on ball reference studs set into the structure. The unit is assembled at site to the appropriate gauge length, inserted between pairs of ball reference studs, and the micrometer adjusted until the end pieces make contact with the balls. Distance readings are then shown on the micrometer barrel.
See also **Portable clinometer**, **Portable extensometer** and **Reference studs**.

Portable shear meter
See **Torvane**.

Portland pozzolana cement
See under **Sulphate-bearing soil**.

Post-hole auger
A simple hand tool used for putting down holes to a depth of about 5 m in soft soils which will stand unsupported. It may also be used with lining tubes if required. Mechanically operated augers are available and are particularly suitable where a large number of holes are to be made.
See also under **Auger**.

Potentiometric extensometer
An instrument for monitoring soil, and rock movement such as the lateral strain and settlement in, or beneath, earth fill, control of slopes and displacement of structures. The potentiometer transducer consists of a rectilinear resistance in a sealed oil-filled housing. Contacts on the extensometer piston slide along the resistance wire. The position of the piston is measured with a null balance portable read-out calibrated directly in units of displacement.
Hydraulic, pneumatic and vibrating wire extensometers working on the same basic principles can also be used.
See also **Magnetic probe** and **Tape extensometer**.

Potentiometric recorder
A unit, used in conjunction with an analogue indicating system, which can directly produce an XY graph in a laboratory soil test.

Power auger
See **Auger**.

Power rammer
This consists of a single air cooled vertical cylinder containing a piston which is rigidly connected by a rod to a spring loaded tamping foot. It is commonly used to compact the backfill of trenches.
See also **Frog rammer**.

Power shovel
See **Face shovel**.

Pozzolan grout
See under **Grout**.

Pozzolanic cement
See under **Sulphate-bearing soil.**

PRA
American Public Roads Administration.

Prager model
A non-linear rheological model to represent an elastic plastic liquid exhibiting delayed elastic recovery when the load is removed.

Prandtl theory
A theory put forward by L Prandtl for the process of penetration of a hard body into a soft homogeneous mass, such as would be the case of a raft foundation of a structure resting on soil.

Preboring
See **Predrilling**.

Precast concrete crib cofferdam
A framework of precast concrete members joined together in criss-cross fashion to form large pockets. These pockets are then filled with rock and sealed with smaller sized material.
See also **Cofferdam**.

Precast pile
Square, rectangular, hexagonal or circular cross-section, reinforced or prestressed, concrete piles cast in moulds and after curing driven into the ground by a pile driver.
See also **Pile shoes**.

Precast sectional pile
See **Brunspile, Composite, Fuentes** and **Raymond piles.**

Precise settlement gauge
An instrument to accurately monitor long-term soil or ground settlement. It consists of a steel casing connected to a settlement plate such that the differential movement between the plate and a surface reference point can be recorded by a dial gauge.

Preconsolidated soil
Soil which has been subjected to a greater effective pressure than its present overburden pressure.

Preconsolidation load (or **Preconsolidation pressure**)
By drawing the curve of compression of a soil which is compressed in the oedometer, an estimate can be made of the highest load to which it has been subjected in the past. This pressure can be translated into a depth of soil and compared with the existing depth of soil. The difference is the erosion caused by glaciation and other processes.

Preconstruction thawing
The technique of building on a frozen soil base that has been prethawed to an appropriate depth.

Predrainage
Lowering the water table in the vicinity of a proposed large excavation which will extend below the level of the existing natural water table.
See also **Well point.**

Predrilling
Removing the soil in the space to be occupied by a pile.

Preliminary pile
A pile installed before the commencement of the main piling works for the purpose of establishing the suitability of the chosen type of pile and for confirming the design, dimensions and bearing capacity.

Preliminary survey
A fact-finding survey carried out in advance of the actual sub-surface investigation. This survey should establish the soil conditions near to the present site, behaviour of nearby structures, articles written about the area, existing borehole and geophysical records, geological maps of the area etc, etc.

Preloaded soil
See **Overconsolidated soil.**

Prepakt cone
A simple cone-shaped instrument for measuring the viscosity of slurry at site. It is slightly different in shape from the Marsh funnel although used in a similar manner.
See also **Marsh funnel** and **Mecasol cone.**

Pressure
Force per unit area of fluid. It is usually expressed in kN/m^2.
See also **Effective pressure** and **Stress.**

Pressure bulb
See **Bulb of pressure.**

Pressure grouting
See **Grouting.**

Pressure membrane apparatus
An apparatus for measuring soil suction.
A sample of soil is placed on a cellulose membrane supported on a porous bronze disc which is in contact with water at zero atmospheric pressure, and a positive air pressure is applied above the sample. Once the sample water content reaches equilibrium the soil suction is then equal to the applied pressure.
See also **Suction plate.**

Pressuremeter
See **Menard pressuremeter.**

Pressure of contact
See **Contact pressure.**

Pressure pile
A board pile, the concrete of which is placed under compressed air pressure.

Pressure ratio
See **Pore pressure ratio.**

Pressure system
A constant pressure system needed for testing soil samples. It can be provided by:—
(1) Self-compensating mercury system.
(2) Oil dash pot.
(3) Air water cylinder and foot pump.
(4) Compressed air/water.
(5) Electro-hydraulic system.

Pressure transducer
A transducer which converts pressure to a proportional electrical signal. for instance, the transducer can replace the dial gauge measuring cell pressure in a triaxial compression test.

Pressure well
See under **Well.**

Prestressed pile
A precast concrete pile, or sheet pile, reinforced with prestressing wires.

Presumed bearing value
The net loading intensity considered appropriate to the particular type of ground for preliminary design purposes. Values for various types of ground are usually given in the form of a table. The particular value is based either on local experience or on calculation from strength tests or field loading tests using a factor of safety against shear failure.

Primary consolidation
See **Consolidation.**

Primary structure
This refers to the arrangement of the soil particles, which may be further described as single particled, flocculated or dispersed. The arrangement is usually developed during the processes of sedimentation or rock weathering. Various discontinuities may occur subsequent to deposition or formation of the soil and these are referred to as the secondary structure.

Principal planes
For any system of forces acting at a point there will be three mutually perpendicular planes on which the stress is wholly normal. These are the principal planes, and the stresses acting on them are known as the principal stresses σ_1, σ_2 and σ_3.

Principal stresses
See **Principal planes.**

Probe
See **Density, Magnetic** or **Moisture probe** or under **Inclinometer.**

Probing
Forcing or driving a probe, such as a rod, pipe or similar device, into the ground and correlating the resistance to penetration to the type or properties of the soil encountered.

Processing
The various operations of soil stabilisation, including pulverisation, moisture control, addition of stabiliser, mixing, rolling, and covering with a primer.

Proctor compaction test
The British standard compaction test in which a cylindrical mould 1/1000 m^3 volume is filled in three layers, each layer of the soil being compacted by 27 blows of a standard rammer of mass 2.5 kg dropped 300 mm at each blow. The bulk density is obtained by weighing the mould and a small portion of the soil is taken for moisture content determination. From these measurements the dry density is calculated. The test is repeated for a wide range of moisture contents and a curve is plotted showing dry density on a base of moisture content. From this the maximum dry density and the optimum moisture content of the soil is read off.
See also **Abbot compaction test, Maximum dry density, Optimum moisture content** and **Vibrating hammer compaction test.**

Proctor mould
A cylindrical metal mould having an internal diameter of 105 mm, an internal effective height of 115.5 mm and a volume of 1/1000 m^3. The mould is fitted with a detachable baseplate and a removable extension approximately 50 mm high. It is used in Proctor compaction tests.

Proctor needle (or **penetration needle**)
A simple instrument for measuring the resistance of a soil to penetration at a standard rate of 12.7 mm per second. Needles of area varying from 32 to 645 mm^2 are used and a spring balance shows the force needed to push the needle in. Granular soil must be screened to remove coarse material.

Proctor penetrometer
See **Proctor needle.**

Proctor rammer
See **Compaction rammer.**

Profile
Horizontal boards fixed on edge at a datum level outside the foundation dig for a small structure. Also sloping boards to act as a datum for the construction of slopes.

Profile gauge
See **Hydrostatic profile gauge.**

Progressive system
A wellpointing system used for trenchwork. It dewaters the ground before the trench is excavated, creating stable conditions. As each trench length is completed, the well points are extracted and reinstated ahead of excavation.
See also **Ring system.**

Proof load
A load applied to a selected working pile to confirm that it is suitable for the load at the settlement specified. A proof load should not normally exceed 150% of the working load on the pile.

Prop
See **Puncheons.**

Prospecting kit
See **Mackintosh prospector.**

Proving ring
A steel ring used for measuring applied load. The ring is manufactured from a strip of high tensile steel and under the action of an applied load its change in diameter is directly proportional to the applied positive or negative load. The change in diameter is measured by a dial gauge or linear transducer which is rigidly mounted within the proving ring and accurately aligned on the loading axis.

Puddle clay
See **Clay puddle.**

Pug
An old name for puddle clay.

Pulling test
See **Pile pulling test.**

Pull rise curve
See **Pile pulling test.**

Pull shovel
See **Back acter.**

Pulse generator
An instrument used in seismic reflection technique of site investigation.
The procedure requires field equipment consisting of a 6 kg rammer with an attached transducer and proximity switch, a twin-channel tape recorder, a shoulder slung control pack and a geophone.
The rammer is dropped to the ground at a distance of about 0.5 m from the geophone. At a rammer height of 150 mm, a sliding rod proximity device operates a pulse generator which causes the tape recorder to run for 500 ms. At the moment of rammer impact the transducer causes a datum pulse to be applied to one channel of the tape recorder while the other channel records the displacement of the main geophone diaphragm caused by the initial and reflected ground motions.

Pumping in test
An *in situ* test to find the coefficient of permeability of a soil where the underlying bedrock is very deep, or where the permeabilities of different strata are required.
A casing, perforated for a metre at its end, is driven into the ground. At stages during the driving, the rate of flow of pumped water required to maintain a constant head in the casing is determined and the soil's permeability so found.

Pumping out test
See **Cone of depression.**

Pun
To ram soil with a punner so as to compact it.

Puncheons
Vertical struts transmitting the mass of the bracing to the excavated ground surface inside a cofferdam or excavation.

Punner
See **Hand rammer** and **Power rammer**.

Push-in piezometer
See under **Standpipe piezometer**.

Push-in piezometer tip
See under **Piezometer tips**.

Putty chalk
A calcareous mud due to the alteration of chalk. The nature of the alteration is not known. In some cases a drift or head derived from the chalk, filling valley bottoms, develops this character.
The material is produced artificially in soft homogeneous chalk by percussion boring tools.
It is generally of clay-like consistency but in wet conditions runs in excavations much like running sand.

Pycnometer
An instrument used for determining the density of soils. The simplest type of specific gravity bottle is merely a jam jar. It is weighed three times, empty, then full of soil, then full of soil and water. If the density of the soil particles is known, these three weighings will also enable the moisture content of the soil to be calculated.
See also **Density bottle** and **Water displacement vessel**.

Pyroclastic
A name given to deposits formed of volcanic ash and signifying their partly igneous and partly sedimentary origin.

Q

Q-test
See **Quick test**.

Quartering
The reduction in quantity of a large sample of soil by dividing a circular heap, by diameters at right angles, into four approximately equal parts, removing two diagonally opposite quarters, and mixing the two remaining quarters intimately together so as to obtain a truly representative half of the original mass. The process is repeated until a sample of the required size is obtained.

Quick clay
A clay having a sensitivity ratio greater than 15.
See also **Sensitivity ratio**.

Quick cohesion assessment
See under **Rapid assessment of plasticity**.

Quick plasticity assessment
See **Rapid assessment of plasticity.**

Quicksand condition
A quicksand condition may result from either of two causes: (1) the flow of water through a sand which reduces the effective pressure between the particles, or (2) the disturbance of the sand, the voids of which are saturated with water and the particles of which are loosely packed.
See also **Well point.**

Quick silt
A silt which is in an unstable condition due to being saturated by water. It is often referred to as bull's liver by American construction workers.

Quick soil classification
See **Rapid soil classification.**

Quick test
A shear test of a soil sample carried out in a shear box or a triaxial compression apparatus without allowing the sample to drain.
See also **Undrained test.**

R

Radar probing
See **Ground probing radar.**

Radiation cone
See under **Ground probing radar.**

Radioactive testing
See under **Nuclear densometer.**

Radius ratio
See **Co-ordination number.**

Raft foundation
A continuous slab of concrete, generally reinforced, laid over the ground as a foundation for a structure. It is as large as, or slightly larger than, the area of the building which it carries. A buoyant raft is a particular and very expensive raft foundation.
See also **Cellular raft.**

Raft stiffening
See **Stiffened raft.**

Rail crib
A crib made from layers of old railway lines.

Rail grillage
See **Rail crib.**

Rainbird
A sprinkler system having temporary quick-connection water supply pipes. The system is used to water an area of dry soil to increase its moisture content.

Rainfall
The amount of rain as measured in a rain gauge. It is usually expressed as millimetres depth calculated on the area of the funnel.

Rain wash
The movement of surface soil and rock down a slope with the help of rain.

Raised beach
A deposit of shingle, gravel or sand situated above the present level of the sea or lake in which it was formed, indicating a relative change in the levels of land and water.

Raisin cake structure
The random occurrence in a till of boulders.

Raker (or **rake**)
An inclined strut.

Raker pile
A pile driven into the ground, at an angle to the vertical, to support loads having a horizontal component.

Rammer
See **Power rammer**.

Rankine's theory
A state of stress theory of granular soil pressures developed by Rankine. His value for the pressure on a vertical wall retaining soil with a horizontal surface is

$$\frac{1 - \sin\phi}{1 + \sin\phi} \times \text{soil density}$$

for each metre depth of soil retained, where ϕ is the angle of friction of the soil. The value $\frac{1 - \sin\phi}{1 + \sin\phi}$ is called the coefficient of active earth pressure.

Rapid assessment of cohesion
See under **Rapid assessment of plasticity**.

Rapid assessment of plasticity
A quick method of approximately determining the cohesion and plasticity of a soil at site as laid down in CP 2001.
To examine a sample for these characteristics it should first be loosened if necessary, for instance by crushing with the foot or a mallet, and then a handful of the material should be moulded and pressed together in the hands. It may be necessary to add water and to pick out the larger pieces of gravel. A soil shows cohesion when, at a suitable moisture content, its particles stick together to give a relatively firm mass. A soil shows plasticity when, at a suitable moisture content, it can be deformed without rupture, ie without losing cohesion. It will generally be possible to distinguish between the presence of silt and of clay, but not to distinguish the range of the liquid limit.

Rapid soil classification
A quick method of approximately classifying soil by using sight and feel as laid down in CP 2001.

Rate

Coarse and fine soils are distinguished by whether they contain more or less than 50% of material nominally finer than 60 microns. Material of 60 micron size feels harsh but not gritty when rubbed between the fingers, and the particles are at the limit of visibility with the naked eye. The presence of coarser particles will make the material more gritty, while finer particles will make it less harsh. Finer, pure silt-sized material feels characteristically smooth and soapy and only very slightly sticky. It is easier to distinguish between gravels and sands, or gravelly and sandy fine soils, because the size which separates the gravel sizes from the sand sizes (2 mm) is easily visible. Particles of 2 mm size are about the largest which will cling together when moist due to the capillary attraction of water. Well-graded and poorly-graded materials can also be distinguished by visual inspection, but this is harder for sands than for gravels.

Rate of consolidation
See **Consolidation settlement.**

Rate of penetration
The penetration resistance of a pile driven into the ground by use of a vibratory hammer. It is expressed as the distance penetrated per second.
See also **Blow count.**

Rate of secondary consolidation
See **Coefficient of secondary consolidation.**

Rate of settlement
See **Coefficient of consolidation.**

Ratio of area
See **Area ratio.**

Ratio of inside clearance
See **Inside clearance ratio.**

Ratio of overconsolidation
See **Overconsolidation ratio.**

Ratio of pore pressure
See **Pore pressure ratio.**

Ratio of recovery
See **Recovery ratio.**

Raymond penetrometer
A 51 mm outside diameter hollow stem penetrometer, 813 mm long, used in a Raymond standard dynamic penetration test.

Raymond pile
A sectional thin concrete corrugated cylindrical shell pile driven into the ground with a mandrel bearing on the bottom and shoulders of the stepped sections. On reaching the required level the mandrel is removed and the shell filled with concrete.

Raymond standard test
A dynamic penetration test to compare the bearing capacities of soils.

RD
Symbol for relative density.

Reactants
Active constituents in a system in which chemical reactions are taking place; used in chemical grouting of soil.

Reaction system
The arrangement of kentledge, piles, anchors or rafts that provide a resistance against which a pile is tested.

Rebound
The opposite to compression.

Recharging
Pumping water back into a water-bearing stratum.

Recovery ratio
The percentage ratio between the length of sample cores recovered and the length of core drilled on a given run.

Redeposited sample
A made up sample of cohesionless soil for testing in, say, a triaxial compression test.

Redriving test
See **Pile redriving test.**

Reduced level
The height above datum.

Reduction coefficient
A reduction factor which must be applied to the theoretical capacity of a pile. This takes into account various factors such as the disturbance of a soil due to the driving of the pile.

Reference studs
These are hard chrome-plated studs, usually spherical in shape, mounted on a cadmium plated steel body which is cemented into a structure or foundation to act as reference points to monitor superficial movement of the structure. They can be either permanent or have demountable heads. Reference studs are usually installed either in a linear pattern, for example along the crest of a dam or down the fall-line of a slope, or on a triangular grid, for example across fissures and construction joints. The pattern of measurements should identify movements and their magnitude and direction.
See also **Permanent reference point** and **Telltale.**

Reflection shooting
See under **Seismic survey.**

Refraction shooting
See under **Seismic survey.**

Reinforced earth
An association of granular soil and linear elements of reinforcement able to resist high traction stresses.

Relative compaction
The percentage ratio of the dry density of the soil *in situ* to the maximum dry density of that soil, as determined by a specified laboratory compaction test.

Relative density (RD)
A measure of the density of a coarse-particled soil which gives a better impression of its compaction than the void ratio.
The densities of the soil are measured in the laboratory in its loosest possible dry state and in its densest possible state. The relative density of a field sample of the same soil is then as follows:

$$\frac{\text{maximum density} - \text{field density}}{\text{maximum density} - \text{minimum density}}$$

or

$$RD = \frac{e_{max} - e}{e_{max} - e_{min}}$$

where e is the void ratio.

Relative plasticity index
See **Consistency index**.

Relative settlement
Differential settlement.

Relaxation
The phenomenon of some piles to decrease their load capacity after being driven.
See also **Freeze**.

Relief hole
See **Weephole**.

Relief platform
A platform at the land side of a retaining wall to transmit heavy superloads direct to the wall and to prevent these loads reaching the back of the wall as additional earth pressure.
A relief platform may sometimes be carried partly on the wall and partly on piles.

Relief well
A borehole drilled at the toe of an earth dam to relieve high pore water pressures caused by the mass of the dam.

Remoulded sample
A sample of soil which has been first broken up into small pieces and then quickly remoulded into a suitable shape for testing, without a significant change in the soil water content.
If, after undrained load testing, there is a change in water content a correction can be made from a plot of water content against the logarithm of the undrained strength of the remoulded soil.
See also **Sensitivity**.

Remoulding index
The ratio of the load per mm of compression of undisturbed clay to the load per mm of compression of the remoulded clay.
See also **Sensitivity**.

Rendhex pile
A hollow steel pile formed by welding two standard trough-shaped steel sections longitudinally together to form a cross-section of hexagonal shape.

Rendulic plot
A three-dimensional graphical plot in which imposed stress combinations can be represented, and stress paths plotted for any specific soil test sequence.

Reno mattress
See **Gabion**.

Representative sample
See **Quartering** and **Riffle box**.

Residual factor
The proportion of the total slip surface, in a soil slope that has failed, along which its shear strength has fallen from a peak value to the residual value.

Residual shear box
A shear box having reversing switches which enable the residual shear strength of a soil to be determined.

Residual soil
A general term used to refer to the results of weathering where some components of the rock have been removed, leaving what is usually a clay-based deposit.

Residual strength
The shear strength of a soil which, after passing through a peak strength, during a shear test, remains at a constant value.
In clay or dense sand the residual strength is usually lower than the peak shear strength.

Resistance cone
See **Cone penetration test**.

Resistivity survey
See under **Geophysical exploration**.

Resoiling
Levelling the ground after dredging, excavating or open cast mining is finished, and replacing the top soil so as to make the ground suitable for growing crops.

Rest period
The wait of ten to twenty minutes after driving a soil sampler to the required depth before starting the actual separation and withdrawal. The rest period allows full development of adhesion and friction between the soil sample and the sampling tube.

Retaining wall
A wall built to hold back soil or other solid material. Important differences in design exist between free and fixed retaining walls. Fixed retaining walls are supported both at the top and bottom so that they cannot tilt, so that the soil pressure does not fall to the reduced value of the active earth pressure. Free retaining walls, however, can be gravity retaining walls. They need not be designed to bend, but must be able to tilt or slide enough to bring on the reduced value of the active earth pressure. They may be of sheet pile construction. Bridge abutments are usually free retaining walls, except for those which are fixed arches and portal frames. Cantilever walls are ‘free’ and may be

counterforted or buttressed, sometimes with a keel in the base to prevent sliding or a relieving platform at the top to reduce soil pressure due to surcharge.

Retaining wall drain
See **Back drain** and **Weephole.**

Retarder
See under **Polymer stabilisation**.

Reversed filter
A graded filter where each layer of material consists of coarser particles than the layer below it.

Revet
To protect a soil surface with a revetment.

Revetment
A protective covering to a soil slope to prevent scour by water or weather.
The covering can be asphalt, concrete slabs, grass, maritime plants, mattresses, non-retaining wall, pitching and rip-rap etc.

Rheological model
An imaginary model composed of springs, dashpots and friction elements, in various combinations, used to study the behaviour of soil.
For details of the different models see under the following headings:— **Bingham**, **Burger**, **Hookean**, **Kelvin**, **Kepes**, **Maxwell**, **Newtonian**, **Prager**, **Schwedoff**, **Standard linear**, **St Venant**, and **Yield stress**.

RIBA
Royal Institute of British Architects.

Ridge line
See under **Watershed**.

Riffle box
A segmented box through which soil can pass; used in the grading and classification of soils and aggregates.

Rig
A portable apparatus erected on the surface of the ground, or supported in a vessel at sea, from which to drill a borehole.

Rigid pavement
See under **Pavement**.

Rim trench
A trench dug around the perimeter of an excavation to drain water to a sump from where it can be pumped away.

Ring magnet
See under **Magnetic probe extensometer**.

Ring shear apparatus
A machine designed to produce measurements of the decrease of the shearing resistance of soils after failure, such as to determine the factor of safety of earth structures.

Ring system
A wellpointing system which provides a 'dry island' for excavation even where the site is adjacent to rivers or the sea. The system of well points may be laid out as a circle, square, rectangle or any irregular shape and connected to a common header pipe through risers. This creates a localised wall of stable soil. The system can dewater to any depth, using two, three or more vertical stages of well points.
See also **Progressive system**.

Ring type cofferdam
See under **Cofferdam**.

Ripper
See **Rooter**.

Rip-rap
Stones to form a revetment.

Risen pile
See **Pile heave**.

River gravel
Gravel having a rounded shape and found on the bed of an existing river or previous river.

Roadbase
The layer of a pavement which provides the principal support to the surfacing.

Road logger
A self-contained mobile unit which tows a trailer having a radioactive source and sensing device. It produces a continuous record of the density and moisture content of a road or airfield pavement, or other material, over which the mobile unit is driven.

Road settlement point
See **Pavement settlement point**.

Rock fill cofferdam
An enclosure formed by mounds of rock. Such cofferdams are suitable for use where the head of water is high or the velocity of water flow is high and there is a suitable hard rock bottom.
They are constructed mainly of massive rock or boulders. Where rock of sufficient size is not available, a quantity of smaller rocks may be enclosed in wire mesh baskets. Its impermeability is improved by filling the voids between the rocks with smaller material.
See also **Cofferdam**.

Rock flour
Silt for which the plasticity index is close to zero.

Rock mechanics
The study of the properties of rocks.

Roc-saw trencher
A machine designed by the Boring and Tunnelling Company of Texas

specifically for working in sub-zero temperatures to cut trenches in hard abrasive permafrost of the Arctic tundra. The trencher has also proved successful in cutting rock in the hot arid climate of the Middle East.
The trencher consists of a large chain saw hung on the back of a modified caterpillar tractor. It is capable of digging 600 mm wide trenches to a depth of 3.5 metres.

Rod extensometer
An instrument to monitor soil, or rock, movement.
It consists of a rod anchored at the lower end of a borehole or drillhole, passing into a reference tube fixed in the hole collar. Relative movement between the end anchor and the reference tube is accurately measured.
This measurement can either be by a dial gauge placed on top of the rod or by an electric transducer registering at a remote read-out terminal.

Rod fork
See under **Bitch**.

Rod tiller
A steel horizontal handle with a split central cramp, used for controlling and turning the tools when drilling and augering.

Roller
See **Sheepsfoot roller** and **Vibrating roller**.

Rolling diaphragm
See **Bellofram**.

Root
That part of a dam which goes into the ground where the dam joins the hillside.

Rooter (or **ripper**)
A towed tool designed to penetrate deeply into hard soil and break it up into fairly large pieces suitable for loading into scrappers or for handling by bulldozers.
See also **Scarifier**.

Rotary drilling
Using a rapidly rotating drill bit to cut or grind a borehole through clay, sand or rock. The fragments so formed are removed from the hole by circulating water or drilling mud.

Rotational failure
A type of foundation failure where the soil below different parts of the foundation is compressed at different rates to cause the settling foundation to rotate on a horizontal axis.
See also **Break in failure** and **Rotational slide**.

Rotational slide (or **cylindrical slide**)
A failure of a clay slope, which involves slipping of the soil on a curved surface.

R_0 test
A constant volume direct shear test developed by O'Neil for the rapid determination of total and effective strength parameters for soils.
It is a consolidated-undrained shear test in which the development of pore

pressure during constant-volume shear is theoretically prevented by changing the vertical load on the test specimen to hold the vertical dial reading constant.

Rotinoff Pile
A precast concrete reinforced shell pile is formed by driving into the ground precast reinforced tubular sections one metre or more long preceded by a concrete shoe. The shells are driven by means of a mandrel passing through them and resting on the shoe. After the required set is obtained the mandrel is withdrawn, a steel reinforcing cage is inserted and the concrete core cast.

Rounded particles
A description of the rounded shape of some coarse soil particles.
See also **Angular**, **Subangular**, and **Subrounded particles**.

Rowe consolidation cell
A consolidation cell, developed by Professor Rowe at Manchester University, which enables soil specimens of much larger diameter to be tested than is normally possible with conventional oedometer apparatus.
The body of the cell comprises three major components: a base, the cell wall and a cover. The cover includes a flexible diaphragm connected through a steel piston shaft to a consolidation dial gauge. A porous disc is attached to the lower side of the flexible diaphragm to assist drainage. Upper and lower drainage connections are provided, together with a connection for the pressure feed to the cell.
See also **Consolidation cell**.

RSGI
Royal Swedish Geotechnical Institute.

R-test
A consolidated-undrained test.

Rubber balloon densometer
See **Balloon densometer**.

Rubber extension modulus
This refers to the modulus of extension of thin rubber, as used in a rubber membrane surrounding a triaxial compression test soil sample, and is required in order to correct the soil compression test observations.
The correction factor (see under **Membrane**) actually requires a knowledge of the compression modulus, but as this cannot be measured for thin rubber it is necessary to measure the extension modulus which can be assumed to be similar.
A circumferential strip, 25 mm wide, of the membrane is stretched between glass rods and the load per unit extension found from which the modulus of extension can be worked out.
See also **Membrane correction**.

Rubber membrane
See under **Membrane**.

Rubber pestle
A standard pestle, 180 mm long overall and having a rubber head 40 mm in diameter and a minimum depth of 15 mm.

Rubber tyred roller
See **Pneumatic tyred roller.**

Rubber water stop
See **Surface waterbar.**

Rubble drain
A trench filled with graded stones which allow water to flow through it.

Rubble stone footing
An old type of footing constructed of pieces of random size masonry joined together with mortar.

Runner
A vertical member used to support the sides or face of an excavation and progressively driven or lowered as the excavation proceeds, its lower end being kept below the bottom of the excavation.

Running sand
Sand below the natural ground-water level, which is carried into the trenches, trial pits or boreholes by the flow of ground-water as excavation proceeds.

Run off
The quantity of water from rain, snow etc, discharged from a catchment area at a given point over a certain period.

Run off coefficient
The ratio of run off to rainfall.
This will vary depending on such surface characteristics as vegetation, condition of soil and steepness of slopes.
For impervious heavy soil having a 2% slope the run off coefficient would be between 0.40 and 0.65, while for moderately pervious soil having the same slope the coefficient would be between 0.05 and 0.40.

Rupture circle
The circle of stress representing soil failure which just touches the envelope of failure (or rupture) line in a Mohr's circle diagram.

Rupture diagram
See **Mohr's circle of stress.**

Rupture line
See **Envelope of failure.**

S

Safe bearing capacity (q_a)
The load per unit area, allowing for a factor of safety, which the soil can carry.

Safety factor (F)
The ratio of the maximum soil restraining condition to the soil disturbing condition.

Sagging soil
See under **Thaw coefficient.**

Saltings
Grassed areas which are covered by tidal water on some or all high tides.

Sample
A specimen of soil. The kind of sample that should be obtained depends on the purpose for which it is required.
Auger or disturbed samples can be used to identify soil strata, obtain moisture content, specific gravity, soil particle analysis, chemical composition and index properties of a soil. However, if the stress-strain properties or density index of the soil is required then undisturbed soil samples must be obtained.

Sample block
See **Hand carved sample**.

Sample mixer
See **Mixer**.

Sample splitter
The American name for a riffle box.

Sampling
The selection of a representative portion of soil.
See also **Quartering** and **Riffle box**.

Sampling gun
A method of supplying sufficient energy to drive a long coring tube into stiff soil. The method consists in utilising the usual drive mass as a gun barrel and the coring tube as the projectile.

Sampling spoon
See **Soil sampler**.

Sand
A natural sediment consisting of the granular and mainly siliceous products of rock weathering, laid down, after transportation either by wind or by water. The individual particles of a sand deposit are visible to the naked eye, and the material possesses no cohesion when dry. In particle-size analysis material of the sand grade is taken as that between 2 mm and 0.06 mm. A sand between 2 mm and 0.6 mm is classed as coarse. A medium sand lies between 0.6 mm and 0.2 mm, while one which lies between 0.2 mm and 0.06 mm is classed as fine. Sands which consist predominantly of one of these grades are described as uniform. A graded sand consists of a mixture of the varying grades; it is described as well graded or poorly graded according to whether the proportions of the various grades are such as to give a high or low bulk density when the sand is well compacted. It should be noted that other definitions for sand grades exist for special purposes, eg BS 882, 'Concrete aggregates from natural sources', and BS 1198, 'Building sands from natural sources'.

Sand bulking
See under **Bulking**.

Sand cone apparatus
See **Sand pouring cylinder**.

Sand drain
See **Vertical sand drain**.

Sand dune
A low hill of windblown sand, usually on the sea shore. The process of selection by wind sorts the sand into areas of very uniform particle size. Generally, the sand becomes finer with increasing distance from the source.

Sand equivalent shaker
A mechanical shaker for sand equivalent tests. The shaker provides a consistent and repeatable oscillation, thus minimising variation from one operator to another.

Sand equivalent test
A rapid field test to show the relative proportions of clay fines and dusts in granular soils and fine aggregates.

Sand fraction
The fraction of a soil composed of particles between the sizes 2 mm and 0.06 mm.

Sand island caisson
A method of sinking a caisson in the dry, where the initial ground level is below water.
For this a ring of sheet piles is driven into the ground to form a cofferdam, which is then filled with sand to form an island. The caisson is then started on the island and sunk through the sand.

Sand piles
A means of deeply compacting the soil by dropping into it a heavy weight such as a pile-driver ram. The ram makes a hole in the ground into which sand is poured. The ram then drives the sand further in and the process is repeated as required. If carried out with damp concrete the process makes a driven cast in place pile.
See also **Vibroflot**.

Sand pouring cylinder
A sand container mounted above a pouring cone and separated from it by a valve. The cylinder is used in determining the volume of a small hole dug in compacted soil for the determination of the bulk density of a compacted soil.
See also **Balloon densometer** and **Nuclear densometer**.

Sand pump (or **sludger, shell pump**, or **bailer**)
A long tube open at the top and fitted with a flap valve at the bottom. It is lowered into a borehole with a cable drill, or similar rig, to extract soil and cuttings.

Sand replacement test
A test to determine, *in situ*, the dry density of compacted soil.
See also **Balloon densometer** and **Sand pouring cylinder**.

Sand sampler
See **Sand pump**.

Sand wick
A porous stocking of woven polypropylene filled with sand and inserted into a small diameter borehole to act as a vertical drain.
See also **Vertical sand drain**.

Saturated density(γ_{SAT})
The density of soil when the voids are completely filled with water.
It is given by:—

$$\gamma_{SAT} = \frac{(Gs + e)\,\gamma_\omega}{1 + e}$$

where Gs = specific gravity of the soil particles,
γ_ω = density of water,
and e = void ratio.
See also **Density**.

Saturated flow
Laminar flow of liquid through a saturated soil.
See also **Non-saturated flow**.

Saturated unit weight
See **Saturated density**.

Saturation degree
See **Degree of saturation**.

Saturation line
(1) The natural water table. Note: capillary water can rise above water table.
(2) The zero air voids line. A line showing the dry density/moisture content relationship for soil containing no air voids.

Saturation zone
The depth of soil throughout which all fissures etc are filled with water under pressure. The upper level of the water is called the water table, and the water contained within the zone is called ground-water or phreatic water.

Saw
See **Wire saw**.

Scaler parameters
In the physical analysis of soil these paramaters include the particle shape, size, size distribution, porosity and surface area.

Scantling
A term used to denote the breadth and thickness of a piece of timber.

Scarifier
A towed tool consisting of spikes and designed to penetrate up to 100 mm into hard soil to break it up into small lumps.
See also **Rooter**.

Schulze Hardy rule
The ion which is effective in causing flocculation is of opposite charge to that on the colloidal particle.

Schwedoff model
A non-linear rheological model to represent an elastic-plastic liquid.
See also **Rheological model**.

Scotch eyed auger
An auger tool incorporating a steel twist drill and spiral point 50 mm in diameter and 200 mm long. It is used for boring through obstructions.

Scraper
An earthmoving machine capable of digging by taking shallow cuts, loading and hauling over considerable distances, and dumping or spreading its load in thin layers on arrival. This work of digging, hauling and spreading is carried out in one continuous cycle. Scrapers may be either tractor towed or motorised.
See also **Cycle time** and **Travel time**.

Scraper bucket
An apparatus for bringing up disturbed soil samples from a borehole. The tubular-shaped bucket is driven into the borehole and rotated. As it turns, scrapings from the side of the hole collect in the lower compartment of the bucket.

Screen analysis
See **Sieve analysis**.

Screw cylinder
A larger edition of a screw pile.

Screw pile
A spiral blade fixed on a shaft and screwed into the ground.

Screws
A slang name for Caisson disease.

Seasonally frozen soil
Soil that has been frozen from one to two seasons.
See also **Frozen soil**.

Secondary clays
Clays moved from the source of formation by natural means and combined with other substances. They are classed as marine (sea deposits), fluviatile (river deposits) and lacustrine (lake deposits).

Secondary compression index
See under **Coefficient of secondary consolidation**.

Secondary consolidation
See **Coefficient of secondary consolidation**.

Secondary structure
See under **Primary structure**.

Sectional piles
See under the following types of pile: **Brunspile**, **Composite**, **Fuentes** and **Raymond**.

Sediment classification
See **Feret triangle.**

Sedimentary deposits
Deposits that were originally laid down as fine sediments, and which now form the solid deposits of the earth's crust, as distinct from drift deposits. Bedded sands, sandstones, clays, slates and limestones are examples of sedimentary deposits.

Sedimentation
Settlement. The sinking of soil or mineral particles to the bottom of the water which contains them. The larger particles settle faster than the smaller particles of the same shape.
See also **Stokes law.**

Seeding
The sowing of grass seed on the surface of newly constructed earthworks.

Seepage
Percolation of small quantities of water through the soil, or a wall. Seepage loss from a canal or reservoir is expressed as a depth over the surface or over the wetted perimeter per unit of time. Inward seepage is called influent seepage, outward seepage is called effluent seepage.

Seepage analogue
See **Electrical seepage analogue.**

Seepage force
The upward force exerted when water flows through a soil under the influence of the excess head of water. The dissipation of head is caused by friction and, because of the loss of energy, a drag or force is also exerted in the direction of flow.
The seepage force is represented by $-h\gamma_w A$
where h is the excess head,
γ_w is the density of water,
and A is the cross-sectional area of flow.

Seepage line
See **Line of seepage.**

Seepage loss
A loss of water through the bank of a canal or reservoir. It is expressed as mm loss in depth per 24 hours, or as cubic metres loss per square metre of bank and bed.

Seepage surface
A boundary of the soil mass where water leaves the soil and then continues its motion in a thin film along the outer boundary of the soil.
Such a situation may occur at the tail water side of an earth dam, when the free surface intersects the slope of the dam at a point higher than the water level at that side of the dam.

Seepage tank
A glass-sided tank about 1.5 m long, 450 mm high and 200 mm wide and having

various holes through which water inlet and drainage sliding pipes and piezometer tubes can be passed.
Models of earth embankments, sheet piled walls and cofferdams etc can be built inside the tank and water seepage flow lines observed with the aid of dye injected into the upstream water.

Segregation
The movement through a soil, or filter, of its very fine particles due to the passage of water.

Seismic survey
A preliminary site investigation method in which miniature artificial earthquakes are produced by explosives, electric sparks, dropping weights or other energy sources generating shock waves in the ground surface layers. These shock waves are recorded by seismometers placed at increasing distances from the explosion. Two methods are noted, reflection shooting and refraction shooting.
Reflection shooting is based on the principle that the vibrations caused by the energy source are partly reflected at the boundary between two strata of very different densities. Reflection shooting works on the principle that energy waves travel at different speeds in soil and rock of different densities.
By measuring the speed and arrival times of the waves, the depths and inclinations of the strata can be computed.
See also **Geophysical exploration.**

Seismogram
A graphic record of the vibrations recorded by seismometers.

Seismometer
An instrument used in a seismic survey to record vibrations.

Selected fill
See **Fill.**

Self-compensating pressure system
See **Pressure system.**

Semi-confined water
Water in a permeable stratum of soil where one or both of the layers above and below the soil stratum are not completely impermeable, but their permeability is nevertheless very small compared with the permeability of the permeable stratum.
See also **Confined water.**

Semigravity wall
A concrete or brick retaining wall which is more slender than a gravity wall and hence requires a foundation. It also requires steel reinforcing.

Semi-hydraulic fill
See under **Hydraulic fill.**

Sensitivity ratio
A measure of the sensitivity of a clay to remoulding. It is the ratio of the unconfined compressive strength in the undisturbed state to that in the remoulded state. Since clays are plastic, an exact failure point is not always

found in the strength tests, and for greater accuracy the strengths at equal strains may be measured. For British alluvial clays it is from 4 to 8. Clays exceeding 8 are termed extra-sensitive and those over 15 are termed quick.

Septa
See **Claystone**.

Septarian nodules
See **Claystone**.

Serrated shoe
See under **Shell auger**.

Set
(1) The net distance by which a pile penetrates into the ground at each blow of the hammer.
(2) The number of blows required on a pile for the last 25 mm of penetration.

Setting
All boards held in position by one frame of 'timber' or in the case of tucking or piling frames, by two adjacent frames.

Setting out
Putting pegs in the ground to mark out an excavation or to locate the position of the proposed foundation.

Settlement
See **Subsidence**.

Settlement cell
See **Hydraulic overflow settlement cell** and **Pneumatic settlement cell**.

Settlement gauge
See **Precise settlement gauge**.

Settlement plate
A 6 mm thick by 1 metre square steel plate with a threaded stud welded at its centre to which a threaded pipe can be fixed.
The plate is used as a control measure when fill material is being placed over compressible soil.
The plate and upstand pipe are placed on top of the compressible soil so that the top of the pipe protrudes through the fill material. As the depth of fill is increased, further known lengths of pipe can be added.

Settlement rate
See **Coefficient of consolidation**.

SfB
An international classification of building information subjects.

SGI
Royal Swedish Geotechnical Institute.

Shaft
An excavation which may be either vertical or inclined, constructed to give access to underground works.

Shaking test
A rapid way of determining whether a sample of fine-particled soil is a clay or a silt. A wet remoulded sample is held in the palm of the hand and if it is a silt its behaviour is as follows: First the surface of the soil is smoothed with a knife and the soil is shaken by tapping the back of the hand. The soil then begins to glisten with the water, which is pushed out of it as the particles slip past each other into a denser position due to the shaking. If the soil pat is then squeezed the water disappears from the surface and re-enters the soil and the glistening appearance vanishes. Cohesive soils show none of these changes. This property of a silt is called dilatancy.

Shale
A laminated clayey or silty deposit which has been compacted naturally and has developed a fissility along the bedding planes.

Shallow footing
A footing founded at a less depth below ground level than the width of the footing.

Shallow pit
See under **Pit**.

Shallow trench
See under **Trench**.

Shear box
A non-ferrous box split horizontally to form upper and lower halves and used to find the direct shear strength of a soil in the laboratory.
See also **Residual shear box**.

Sheargraph
See **Cohron sheargraph**.

Shear slide
A landslip in which a mass of soil slides as a block away from the material below it.

Shear strength (τ)
See under **Angle of internal friction** and **Angle of shearing resistance**.

Shear strength tests
The shear strength of soil samples is measured in the laboratory mainly by one of the following tests:— For soft cohesive soil by the laboratory vane test, for cohesive soil by the triaxial compression test and for cohensionless soil by the shear box test. At site the approximate shear strength of cohesive soil samples can be measured using the Pilcon vane, or Torvane, or by testing 38 mm diameter by 75 mm long cylindrical samples in an unconfined compression test apparatus. However, since the behaviour of soils *in situ* often differs from their laboratory behaviour, the *in situ* vane apparatus can be used to find the shear strength of cohesive soil at site.

Sheepsfoot roller
A towed roller for compacting silty or clayey soils. It consists of one or more large drums which can be water ballasted and which have protecting studs.
See also **Tamping roller**.

Sheeted pit
An excavation in which the sides have been sheeted.

Sheeting
Vertical planks driven into the ground adjacent to the walls of a trench or excavated hole to prevent the walls collapsing.

Sheet pile cofferdam
See under **Cofferdam**.

Sheet pile cut off
See **Cut off wall**.

Sheet piles
Closely set piles of timber, reinforced or prestressed concrete, or steel, driven vertically into the ground in a line to keep soil or water out of an excavation.

Sheet pile wall (or **sheet piling**)
A wall of sheet piles, which may be a cantilever wall or anchored back at one or two levels.

Shelby tube sampler
A thin-wall seamless tube sampler which is driven into the bottom of a borehole, or trial pit, to obtain an undisturbed soil sample. These samplers are usually 750 mm long and vary in internal diameter from 45 mm to 100 mm.

Shelf
See **Continental shelf**.

Shelf retaining wall
A reinforced concrete retaining wall with a relieving platform built on to its upper part.

Shell auger
A tubular auger used for cleaning cuttings and sediment from a borehole following drilling by chisel or bits. The auger base of the shell can also be unscrewed and replaced with a 'gravel' shoe for penetrating gravelly beds and stony formations, or with a serrated shoe for cutting into soil as the shell is rotated.

Shell marl
A white or pale buff coloured deposit formed largely of fresh water shells and calcium carbonate, usually soft and easily dug. It may be regarded as an unconsolidated limestone of recent formation. There is usually a proportion of clay or silt present.

Shell-perm process
Injection of bitumen emulsion into a permeable soil to reduce the flow of water into an excavation. The emulsion contains a coagulator to make it solidify in the ground. This effectively closes the pores but does not strengthen the ground.

Shell pile
See **Armco**, **Cobi**, **Monotube** and **Raymond piles**.

Shell pump
A sand pump for bailing out boreholes.
See also **Sand pump**.

Shield
A steel cover which protects men driving a tunnel through soft soil.

Shoe
See **Cutting shoe** and **Pile shoe**.

Shore
A prop or support for the side of a building.

Shoring
The provision of temporary supports to a structure.

Short-bored piling
Bored piling about 3 to 5 m deep used to support domestic property on poor strength soil. Fixed over the strategically placed piles are beams to support the walls and floor of the property.

Shrinkage
(1) The reduction in volume of a soil caused by the removal of contained moisture. This could be due to tree roots sucking out the soil moisture during the growing season or in times of drought, or to a building being constructed over the soil so that the natural moisture content is reduced and the soil dries.
(2) The reduction in height of recently laid fill material, such as in an earth embankment, backfill behind a retaining wall or in an excavation.

Shrinkage degree
See **Degree of shrinkage**.

Shrinkage limit (SL)
The maximum water content for a given soil at which a reduction in water content will not decrease the volume of the sample. It describes the limit between the solid and the plastic states of a clay and is usually distinguished by a colour change, the clay becoming much paler at water contents below the shrinkage limit.
See also **Atterburg limits**.

Shrinkage limit apparatus
An apparatus to measure the shrinkage limit of a cohesive soil. It consists of an enclosed bath of mercury and a hanging cage, for the soil sample, which can be lowered into the mercury bath through a hole in the top. The initial volume of the saturated soil sample is measured by immersing it into the mercury bath and measuring the total volume of mercury, cage and soil sample. The cage is then raised so that the soil air dries, after which the case is lowered into the bath and the new reduced volume measured. This process is repeated until no further reduction in volume of the soil sample takes place.

Shrinkage limit test
See under **Shrinkage limit apparatus**.

Shrinkage ratio (SR)
The rate of decrease per unit volume with decrease in moisture content.

Side boards
Boards forming the sides of a heading.

Side drain
A non-softening filter paper which has wide vertical slots cut away from it. It is used to wrap round a cylindrical soil sample of low permeability to improve drainage during the consolidation stage of a drained test.

Side scan sonar
A new technique to quickly and accurately survey large areas of reservoir bed and dam embankment without dewatering.
Two sonic transducers fixed to each side of a metallic fish towed by a boat sweep backwards and forwards covering the survey area. One transmits signals and the other relays the response from the reservoir bed, or embankment wall, to a recorder in the boat. The navigational system comprises two radio transmitters at known positions on the shore and a 'pick up' in the boat. By careful integration these two elements produce a two-dimensional topographical representation of the bed or wall in the form of a photographic negative which can also clearly show the position of holes and cracks in the soil.

Side trees
'Timbers' which support the head trees and side boards in a heading.

Sieve
A circular vessel with a bottom of woven wire to separate the fine soil particles from the coarse particles and manufactured in accordance with British standards.
See also **Flakiness sieves.**

Sieve analysis test
A test in which a sample of air-dried soil is shaken through a series of reducing size mesh BS sieves in order to determine the particle size distribution of the soil sample.
See also **Particle size distribution chart.**

Sieve shaker
A mechanically vibrated table which has two vertical rods and an adjustable top retaining plate which clamps on to a bank of sieves. The apparatus usually has a time switch which allows the sieves to be shaken for any length of time up to 60 minutes.

Sight rail
A horizontal rail set at a specific height above a required level, such as the formation level of an excavation.

Signal conditioning module
A unit, forming part of an analogue or digital indicating system, which provides the transducer excitation voltage and accepts the incoming signal from the transducer. For instance, a radio or television set is a signal conditioning module.

Signal man
The American term for a banksman.

Silent piling
A system of driving piles into the ground under hydraulic pressure. It is

particularly useful where vibration or noise cannot be permitted, such as near hospitals.
See also **Muffler piling.**

Silicate injection
See **Joosten process.**

Sill
A 'timber' laid across the bottom of a heading or trench and carrying at its ends the feet of the side trees. A liner or chogs may be used on the sill to keep the side trees or ground props apart.

Silt dilatancy
See under **Shaking test** and under **Silt.**

Silt fraction
The fraction of a soil composed of particles between the sizes 0.06 mm and 0.002 mm.

Silt grade
Material of silt size.

Silts
A natural sediment of material of finer grades than sands, consisting of granular and mainly siliceous products of rock weathering. The individual particles are only barely distinguishable by the naked eye, but the material has a gritty touch when squeezed between the finger and thumb. It can be rolled into threads between the fingers but crumbles readily when it dries. If a pat of wet silt is shaken in the hand, water will be exuded and, if the pat is then pressed, this water retreats into the silt, leaving a matt surface, because of an increase in volume. This property is known as dilatancy. A silt consists of particles between 0.06 mm and 0.002 mm which possesses a small degree of cohesion and plasticity.

Single particled soil
A description of a soil where each particle touches several of its neighbours in such a way that the whole is stable even if there are no forces of adhesion at the points of contact between the particles.

Single point mercury settlement gauge
See **Pneumatic settlement cell.**

Single shot grout
See under **Grout.**

Single skin cofferdam
See under **Cofferdam.**

Sinker bar
A heavy steel tube about 1¼ metres long which fits over a boring rod immediately above any boring tool to which it is required to add mass to assist in soil penetration.

Sink hole
See **Swallow hole.**

Site
An area of land which is to be used for a construction project.

Site investigation
The examination of the surface and subsoil at a site to obtain the information needed for the design of the foundations and therefore of the remainder of the structure. The most expensive part of this work is the sinking of boreholes, test pits, shafts, or headings under the structure. Exploration usually takes place to a depth equal to 1.5 times the width of the building.
See also **Borehole log**.

Skeletal diagram
An upright rectangular diagram horizontally split into sections to represent, in the correct proportions, the solid, liquid (oil or water) and gas (usually air) portions of a particular soil sample.

Skempton, Alec *(born 1914)*
Professor of Civil Engineering at Imperial College London from 1946. Founder of the concept of the pore pressure coefficients.

Skempton coefficients
See **Pore pressure parameters**.

Skimmer
A bucket which slides backwards and forwards along a boom and used for excavating material slightly above, at the level of, or slightly below the level of the tracks of the machine to which it is mounted.

Skimming
Removing the irregularities of the surface of the soil.

Skin friction
The resistance of the soil surrounding a pile or caisson to its movement. It is usually proportional to the area of pile or caisson in contact with the soil and its value also increases with the depth of penetration.
See also **Adhesion factor** and **Negative skin friction**.

Skip
A hoisting bucket used in deep excavating.

Slab
(1) A term used to define that part of an airfield or road pavement concrete enclosed between vertical joints.
(2) Plain or reinforced concrete which may combine the functions of the running surface of a road or airfield pavement or serve solely as the base for a bituminous surfacing.
(3) A plain or reinforced concrete flat slab which is constructed directly onto a prepared soil or hardcore base to act as both structural foundation and ground floor of a building.

Slaking
When an air-dried or oven-dried sample of fine-particled soil is placed in water the soil slakes. The force of surface tension draws the water into the voids and compresses the air trapped inside. The pressure in the air may become so high

that the corresponding tension in the soil skeleton exceeds the tensile strength of the soil and causes the slaking.

Sleech
A name used, particularly in Northern Ireland and Scotland, for recent alluvial silt.

Sleeper
See **Foot block**.

Sleeve grouting
A method of grouting using a sleeve device known as a 'tube-a-manchette'. It consists of boring a 100 mm diameter hole through the ground and then placing within it a smaller steel pipe known as a tube-a-manchette having groups of peripheral holes spaced 300 mm vertically apart. These holes are covered with a rubber sleeve which acts as a valve. The annulus between the pipe and the borehole is then filled with a plastic type of material which is allowed to set. Grout is introduced through a 20 mm diameter pipe, the end of which has packers on either side of orifices which permit the grout to be discharged through any selected sleeve in the tube-a-manchette. The grout then forces its way through the peripheral material into the adjoining soil.
See also **Grouting**.

Slickensides
Polished surfaces in stiff clays that have had differential movement or expansion.

Sliding resistance
The Shear strength between dissimilar materials such as soil and concrete.

Sliding-wedge theory
The wedge theory for determining graphically or by calculation the passive or active earth forces on a retaining wall. When the properties of the soil are known from laboratory tests on undisturbed samples this method is preferable, particularly for clays, to the use of Rankine's theory.

Slight plasticity soil
A soil having a plasticity index of between one and five.
See also **Plasticity index**.

Slip
A small landslip.

Slip circle
The assumed circular arc of failure of a clay bank.

Slip surface
The surface of failure of a soil bank.
See also **Rotational slide**.

Slope angle
The angle of a soil slope. It can either be given in degrees to the horizontal, or as a tangent of the angle to the horizontal (eg a slope of 1 in 3).

Slope meter
A probe-type instrument, which embodies a suspended pendulum mass, for measuring the angular displacement of soil in an embankment or cutting.

Slope stability
The study of soil slopes which form cuttings, embankments and earth dams.

Slotted sinker bar
See under **Sinker bar.**

Slow test
A drained shear test.

Sludge pump
A pump used to extract water containing a high proportion of solids from an excavation.

Sludger
See **Sand pump.**

Sluiced fill
See under **Hydraulic fill**

Slurry
As used in diaphragm wall construction is a bentonite-water mix. It normally has a base-exchanged bentonite concentration between 4% and 7%.

Slurry sampler
An apparatus for obtaining a sample of slurry from any depth within a diaphragm wall trench.
It consists of a sample tube with a ball valve at its base, which is operable from the surface.

Slurry trench
A trench filled with bentonite slurry and used as an aid in the construction of diaphragm walls.

Slush pit
A long pit excavated adjoining a drilling hole.
To remove debris resulting from drilling, while at the same time maintaining the liquidity and density of the bentonite slurry, the slush pit is excavated with the end adjacent to the hole deeper than the other end. The capacity of the pit should be at least twice the ultimate depth of the drilled hole.
The loaded slurry is raised up the hole and then pumped to the shallow end of the slush pit. As the slurry flows through the pit, settlement of the debris occurs, and the remaining mud flows back into the hole after passing over a weir.

Smear
In the installation of sand drains the soil in a zone next to the 'well' is remoulded, usually causing a large reduction in the permeability of the disturbed zone. This condition is referred to as smear.
As an approximation, the effect of smear is sometimes taken as equivalent to halving the 'well' radius.

Smectite
A crystalline hydrous alumino-silicate mineral of clay.
See also **Illite** and **Kaolinite.**

Smell
See **Odour**.

Soakaway
An excavated hole which can be empty, or filled with rubble, into which surface water is drained to gradually soak away into the soil.

Sodium hexa meta phosphate (Na $PO_3)_n$
The most commonly used dispersing agent.

Sodium oxalate ($Na_2C_2O_4$)
See under **Dispersing agent**.

Soft
A cohesive soil is said to be soft when its undrained shear strength lies between 15 and 40 kN/m^2.

Soil
In the engineering sense, any naturally occurring loose or soft deposits forming part of the earth's crust, particularly where they occur close enough to the surface of the ground to be encountered in engineering works, but excluding topsoil. The term covers such deposits as gravels, sands, silts, clays and peats. It should not be confused with the agricultural or pedological soil which embraces only the topsoil and subsoil. Pedological soil may come within the meaning of the word soil in the engineering sense when it is excessively deep as in some tropical and continental regions, but in the British Isles the two conceptions of the word are distinct.

Soil auger
See **Auger**.

Soil-cement cylinders test
A steel mould having an internal diameter of 71 mm by 229 mm high which contains upper and lower pistons to produce a standard size soil-cement sample of 71 mm diameter by 142 mm high for testing.

Soil classification
See **Airfield soil classification**.

Soil classification assessment
See **Rapid soil classification**.

Soil conditions
See under **Consistent condition**, **Erratic condition**, **Stable condition** and **Unstable condition**.

Soil creep
See **Creep**.

Soil expansion
The increase in volume of a soil sample, when it is removed from a sample tube, without a change in water content.

This may be caused not only by air entrapped in the pores but also by air and other gases which are dissolved in the pore water and are released and expand when the pore water pressures and soil stresses are decreased.

Soil heave
See **Frost heave, Heave** and **Pile heave.**

Soiling
See **Resoiling.**

Soil lathe
See **Lathe.**

Soil legend
The conventional graphic symbols as laid down in CP 2001 for showing on paper the soil type, such as in a borehole log.

Soil map
A map showing engineering soil patterns. It is produced mainly from an interpretation of aerial photographs backed up with geological information, field reconnaissance and borehole soil samples.

Soil mechanics
The investigation of the composition of the soils, their classification, consolidation, strength, the flow of water through them, and the active and passive earth pressures in them. The science was christened under this name at the first International Conference of Soil Mechanics and Foundation Engineering at Cambridge, Massachusetts in 1936.

Soil mixer
A machine used for pulverising the soil in soil stabilisation. The term includes plant mixers as well as travel mixers.
See also **Mixer.**

Soil name
The name given to different types of soil and based on particle size distribution and plastic properties. These characteristics are used because they can be measured readily with reasonable precision and estimated with sufficient accuracy for descriptive purposes, and because they give a general indication of the probable engineering characteristics of the soil.

Soil phases
Any homogeneous part of a soil system different from other parts of the system and separated from them by abrupt transition, ie the solid soil particles and the water in the voids.
Soil consists of fundamental phases as below:—

(1) Solid phase, which consists of the mineral particles and organic material.
(2) Liquid phase, which consists of the water or oil filling all or part of the voids.
(3) Gaseous phase, which consists of the water vapour or gas entrapped in the part of the soil voids not occupied by water or ice.
(4) Ice phase, which consists of ice filling all or part of the soil voids.

Soil plasticity chart
See **Plasticity chart.**

Soil profile
A vertical section showing the succession of soils at a site.

Soil sample
See **Sample.**

Soil sampler (or **clay sampler** or **sampling spoon**)
A tube which is driven into the ground with the object of obtaining an undisturbed soil sample. They are used mainly for cohesive soils.
See also **Coredrill, Core lifter, Hollow stem auger, Liner** and **Split spoon sampler.**

Soil saw
See **Wire saw.**

Soil shredder
A machine, used in soil stabilisation, consisting of two half drums which just do not touch, rotate in opposite directions, and break up the soil.

Soil shrinkage
See **Shrinkage.**

Soil solidification
See **Freezing process, Grouting, Polymer stabilisation** and **Water glass injection.**

Soil stabilisation
Any artificial method of increasing the cohesion and hence strength of a soil. Most methods involve the injection of slurries or chemical solutions into the voids of a soil.
These materials—such as cement grout, clay slurry, water-glass, hydrated lime—then harden to varying degrees and thus strengthen the soil. Since these materials also partly fill the soil voids the permeability of the soil is also reduced.
See also **Freezing process.**

Soil strain gauge
An instrument developed by the TRRL for the measurement of dynamic and long-term strains in soils beneath road pavements.
The strain gauge consists of two aluminium end plates, 64 mm in diameter, attached to a differential transformer. The two coils of the transformer are wired to form a 'half bridge' circuit and the change of inductance of the coils as the core moves is measured.

Soil suction
The negative pressure by which water is retained in the voids of a soil sample when the sample is free from external stress.
See also **Matrix, Osmotic suction, pF scale** and **Total suction.**

Soil survey
The thorough examination of the soil at a site and recorded in a final report. The report will include the details and analysis of all preliminary explorations, field tests, boreholes sunk and the results of comprehensive laboratory testing of the soil samples obtained. It may also include recommendations for specific types of foundations to be used at the site.

Soil symbols
See **Soil legend**.

Soil uplift
See **Heave**.

Soldier
A vertical member supporting horizontal poling boards or walings across an excavation.

Soldier piles
Piles driven into the ground adjacent to the edge of a proposed excavation. Depending on the type of soil they are spaced between 1 and 3 metres apart. As the soil next to the piles is removed, horizontal boards are introduced and wedged against the soil behind the piles. As the depth of excavation increases wales and struts can be inserted in the manner as for sheet piling.

Soletanche system
A pressure grouting system of French origin.

Solidification of soil
See **Soil solidification**.

Solid phase
See under **Soil phases**.

Solum
A term used in Scotland for the area within containing walls of a building, after removal of the top-soil and vegetation.

Solute suction
The difference in pressure between pure water permeating through one side of a semi-permeable membrane and a solution on the other side.
This suction may be as high as one atmosphere (pF = 3) between the layers of a montmorillonite crystal, but rarely exceeds pF = 2 in the free water between the particles.

Solution channel
An underground channel formed by the passage of water.
This could be caused by a continuous flow of rainwater passing down through limestone, or by subsurface water coming up to the surface. Eventually the overlying soil caves in.

Solution channel
See under **Grout**.

Sonar scan
See **Side sonar scan**.

Sounding
See **Probing**.

Spall drain
See **Rubble drain**.

Specific discharge
See **Discharge velocity**.

Specific gravity (G_s)
The specific gravity of soil particles is the ratio of their density to that of water. See also **Pycnometer**.

Specific gravity bottle
See **Density bottle**.

Specific surface
The surface area of unit weight of soil particles or other material.

Specific weight
The density of a material multiplied by the acceleration of gravity g.

Speedy moisture meter
An apparatus for the rapid measurement of the moisture content of fine soils and sand.
A known mass of the wet soil is placed in a pressure container and a quantity of calcium carbide is added. The container is then sealed and the two materials brought into intimate contact by hand shaking. The reaction of the carbide on the water in the soil sample produces a specific quantity of acetylene gas. A pressure gauge, calibrated for moisture content, is fitted at one end of the cylinder and this is actuated by the quantity of gas produced.

Spill through abutment
This consists of two or more columns carrying a beam that supports the bridge. The fill extends on its natural slope from behind the bridge beam through the openings between the columns. In its extreme form it consists of a row of piles driven through the fill and supporting the bridge.

Split barrel sampler
See **Split spoon sampler**.

Split former
A tube which longitudinally splits into halves or thirds and used to prepare soil samples of specific dimensions.

Split spoon sampler
A 35 mm internal diameter sampling tube, 600 mm long, which can be split longitudinally to reveal the undisturbed soil sample obtained from the bottom of a borehole or trial pit.
When driven into the soil with a 63.5 kg mass falling 762 mm, the blows per 300 mm required to drive the sampling spoon are a measure of the standard penetration resistance. The operation is known as the standard penetration test.

Spoil
See **Waste**.

Spongy
Refers to an open-structured soil which is very compressible.

Spoon
See **Soil sampler**.

Spoon auger
A tube auger which is split along the length of one side. When it is rotated in a borehole its side cuts into the soil.

Spot level
The elevation of a point above datum level.

Spotter
The American term for a banksman.

Spotting
In soil stabilisation, laying bags of stabiliser in position on the ground to be stabilised, at regular intervals.

Spread
The arrangement of seismometers in relation to the shot point in a seismic survey.

Spreader
A machine, mounted on wheels, for distributing loose material while in transit.

Spread footing
A footing that supports a single column.

Spring
A place where ground water flows out naturally on the ground surface.
See also, **Fault spring**, **Stratum spring** and **Valley spring**.

Sprinkler
See **Rainbird**.

Sprung foundation
A foundation incorporating springs and designed to support machinery producing dynamic loading or having excessive vibration.

SPT
Standard penetration test.

Spudding
Driving spuds, which consist of hard metal points, into a stratum of stiff clay or soft rock which lies at a shallow depth. This breaks up the soil or rock, after which the spuds are withdrawn and piles driven.

Spud in
To commence drilling a borehole by 'making a hole'.

Stabilisation of soil
See **Soil stabilisation**.

Stabilised overburden pressure
Overburden pressure, existing at a given point in a soil mass, which has reached a condition of stability with respect to the supporting bearing stratum.

Stabilised power module
A unit which provides a power supply at a stabilised voltage to a signal conditioning module.

Stabiliser
Any material added to a soil in soil stabilisation. It may be a chemical which absorbs water, a waterproofer such as bitumen, a cement or a resin.

Stability number (N)
A non-dimensional number $N = C/F\gamma H$ used in conjunction with Taylor's curves,
where C is the soil cohesion,
F the factor of safety,
γ the soil density
and H the vertical height of a slope.
See also **Taylor's curves**.

Stability of slopes
See **Slope stability**.

Stable condition
A term used in tunnelling where the materials or soils encountered stand firmly throughout the overall excavation.
See also **Unstable condition.**

Staging
A working platform supported on the main framing of trenches.

Standard linear model
A complex composite rheological model to represent a complex viscoelastic response of soil.
See also **Rheological model**.

Standard load test
See **Plate bearing test**.

Standard penetration test
See **Penetration tests** and **Split spoon sampler**.

Standard pile
See **Guide pile**.

Standing-water level
See **Ground water**.

Standpipe piezometer
An instrument to measure and control water pressure in soil. It consists of a 25 mm internal diameter standpipe tube with a porous piezometer tip connected at its lower end. It is installed in a borehole and bentonite and grout are used to seal the borehole above the tip, which is surrounded by a sand filter. Groundwater can thus enter the tube only via the tip. The soil water pressure is measured with a dipmeter water level probe and corresponds to the height of the water in the standpipe above the piezometric tip.
Alternative types may be driven or pushed into soft soil and different types of tip are used to suit various types of soil.

Static penetration tests
Soil tests in which the testing device is pushed into the soil with a measurable force, as opposed to dynamic penetration tests in which the testing device is driven in by blows from a standard hammer. Plate bearing tests, the penetrometer, and the cone penetration tests are examples, possibly also the vane test and the Proctor plasticity needle.

Stay pile
A pile driven or cast in the ground as an anchorage for a land tie holding back a sheet-pile wall, etc.

Steady flow
The flow of a liquid, the characteristics (density, pressure and velocity) of which are independent of position, temperature and time.

Steel heave gauge
See **Heave gauge.**

Steel pile
In soft soil a steel tube, usually filled with concrete after being driven, or in hard soil an H-section universal column.

Steel pipe pile
See **Pipe pile.**

Steel sheet cofferdam
See under **Cofferdam.**

Steel sheet piling
Sheet piling of interlocking rolled-steel sections driven vertically into the ground along the edge of a guide waling before excavation is begun. When the sheet piling is completed the excavation can be begun in safety. It keeps out flowing ground, often also water, but requires heavy strutting against either of these, unless its penetration depth below the lowest dig level is relied upon to support it. In this case the penetration depth is about equal to (or more than) the length of piling above dig level; also a very strong piling must be used. Strutting or tying back saves steel, and is therefore used where possible.
See also **Bulkhead.**

Stem
The vertical part of a reinforced concrete retaining wall.

Stemming
To prevent the loss of soft or loose soil, such as dry sand, through joints or openings in 'timbering'. Straw, cement bags or similar material may be pushed into the crevice to stop or 'stem' the flow.

Stepped foundation
A benched foundation.

Step taper pile
See **Raymond pile.**

S-test
See **Slow test.**

Sticky limit
The water content of a soil which ceases to stick to metal.

Stiff
A clay is said to be stiff when its undrained shear strength is between 100 and 150 kN/m^2. It will require a pick or preumatic spade for its removal.
See also under **Clay.**

Stiffened raft
A reinforced concrete raft foundation stiffened to prevent excessive deformation. The stiffening can take many forms, such as over-reinforcing or utilising the stiffness of the supported structure.

Stoke's law
A law of settlement which states that small spheres in a liquid settle at different rates according to the size of the sphere.

Stone column technique
A process used, since the late 1950's, for strengthening cohesive deposits such as alluvial clays and silts, and in these circumstances it is widely used for the support of oil storage tanks.
The process consists of using a cylindrical poker which carries inside its bottom section an eccentric mass rotated by either electric or hydraulic motor. The poker, or vibroflot, penetrates the ground by vibration, assisted in certain conditions by water jet from the nose cone. One reaching the desired depth the vibroflot is raised slowly, allowing the horizontal centrifugal force developed by the rotating eccentric mass to produce the required densification of imported coarse gravel backfill to form a dense stone column.

Stone drain
See **Rubble drain.**

Stone footing
See under **Dimension stone footing** and **Rubble stone footing.**

Stop end tubes
Smooth, clean steel tubes placed vertically into each end of a diaphragm wall trench before concreting begins. They are fixed into the bottom of the trench to stop concrete rising up their inside. After the concrete has set the tubes are removed and the semi-circular groove left acts as a key for the next section of diaphragm wall.

Storm water
Rain-water collected by the drains of a building and the adjacent ground.

Strain
A non-dimensional ratio, which in linear terms is expressed as:—

$$\frac{\text{change in length}}{\text{original length}}$$

Instruments used to measure soil strain include extensometers and strain gauges operating on electrical, mechanical and optical principles.

Strain gauge load cell
See **Load cell.**

Strap footing
See **Strip footing.**

Strata
Plural of stratum.

Stratum
A bed of soil formed by natural causes and consisting usually of a series of layers.

Stratum spring
Sometimes called a 'contact spring', it is caused when water passing downward through a permeable stratum is held up by an impervious layer.
See also **Fault spring, Spring** and **Valley spring.**

Streaming potential
The induced electric potential difference in a thermo-osmotic process. It may be considered as the converse phenomenon of electro-osmosis.

Strength envelope
See **Envelope of failure.**

Stress
Force per unit area of solid. It is usually expressed in kN/m^2.
See also **Effective stress** and **Pressure.**

Stress circle
See **Mohr's circle of stress.**

Stress distribution
See **Principal planes.**

Stress strain curve
A graph showing the relationship of the stress experienced in a soil against the respective values of strain.

Stretcher
See **Liner.**

Strip footing (or **strip foundation**)
A foundation for a wall or for a succession of closely spaced columns or piers. It is ribbon-shaped, projects slightly each side of the wall, and may be of brick-work, timber (if permanently submerged), mass-concrete, or reinforced concrete.

Stripping
Clearing a site of top-soil in order to start construction.
See also **Clearing** and **Grubbing.**

Strip turfing
See **Turfing.**

Strut
A member in compression, eg supporting the walings of a cofferdam or trench.

Stub wings
Small wing walls near the top of an abutment, which allow the fill to spill out around the abutment.

St. Venant model
A simple composite rheological model to represent the elastoplastic response of soil.
See also **Rheological model.**

Subangular particles
A description of the shape of some coarse soil particles.
See also **Angular particles, Rounded particles** and **Subrounded particles**.

Sub-artesian
See **Artesian water**.

Sub-base
A bed of material laid under a roadbase on the natural ground to strengthen it, to improve the drainage, or for some other purpose.
See also **Pavement**.

Sub-drains
Open-jointed or perforated pipes laid in a trench at the bottom of excavations to drain the ground as the work proceeds.

Subgrade
The upper part of the soil, natural or constructed, which supports the loads transmitted by the overlying pavement.

Subgrade reaction modulus
See **Modulus of subgrade reaction**.

Subgrade restraint
Frictional resistance offered by the subgrade to the horizontal movement of a pavement slab.

Sub-irrigation
Raising the ground-water level near the roots of plants etc.

Submarine sampling
Carrying out soil sampling through open water in bays, lakes, ocean shores and rivers.

Submerged density (γ^1)
The density of a soil below the water table. It is equal to its bulk density less the density of water, ie $\gamma^1 = \gamma - \gamma_w$.

Submerged permeable boundary
A submerged liquid/soil boundary through which the liquid can permeate, such as the face of a water-retaining earth embankment where water seepage takes place.

Submersible pump
A compressed air or electrically operated centrifugal pump which can operate when wholly submerged in water.

Subrounded particles
A description of the shape of some coarse soil particles.
See also **Angular**, **Rounded** and **Subangular particles**.

Subsidence
The downward movement of ground. It may be due to a variety of causes, such as underground workings, changes in the ground-water table, differential settlement due to poor foundation design, or mass ground movement associated with landslides or superficial creep.

Subsoil
The weathered portion of the earth's crust that lies between the topsoil and the unweathered material below. The term is sometimes used to refer to the soil in the engineering sense, but its use in this sense is deprecated.

Subsoil drain
See **Agricultural drain**.

Substructure
That part of any structure which is below natural or artificial ground level, in particular the foundations and piers of a bridge.

Subsurface erosion
Piping.

Subsurface water
The term used to define all water found beneath the earth's surface.

Suction
See **Soil suction**.

Suction plate
An apparatus for measuring soil suction.
A weighed sample of soil is placed on the upper surface of a fine porous disc of sintered glass, while the underside of the disc is in contact with water to which a suction pressure is applied. When the water content of the soil sample reaches equilibrium with the applied suction pressure, the sample is reweighed.
See also **Pressure membrane apparatus**.

Suction plate apparatus
An apparatus to measure soil suction.
It consists of a glass cylinder which can be sealed at its upper end and is connected via a tube from its lower end to a vacuum pump.
To measure the soil suction a specimen of soil is placed on the upper surface of a fine porous disc of sintered glass which floats, as a piston, on the surface of the water. The cylinder is then sealed and a suction pressure is applied to the water and a transfer of moisture occurs until the disc and specimen are in suction and vapour equilibrium.

Suitable fill
See **Fill**.

Sulphate-bearing soil
If ground-water contains more than 0.1% of SO_3 or if a clay contains more than 0.5% of SO_3, sulphate-resisting cement should be used for all concrete in the ground. Portland pozzolanic cement may sometimes give enough protection at lower cost. No precautions need be taken with foundation concrete in water which contains less than 0.02% SO_3 or clay which contains less than 0.1%.

Sulphate resisting cement
See under **Sulphate-bearing soil**.

Sulphate test
See **Barium sulphate test**.

Sump
A pit dug below the bottom level of an excavation to collect water.

Supercooled soil
Soil that contains strongly bound or mineralised water which does not freeze at the particular subfreezing temperature.
See also **Frosted soil, Frozen soil** and **Thawed soil**.

Superficial compaction
The compaction of soils by the frog rammer, hand punning, vibration, sheepsfoot rollers, pneumatic-tyred rollers, or similar methods, in layers usually not exceeding 150 mm.

Superimposed load
A specific value of live load.

Superload
See **Superimposed load**.

Surcharge
Any load above the soil which is level with the top of a retaining wall. Surcharges may be temporary (live) loads such as lorries, locomotives, cranes or stacked material; or permanent (dead) loads such as soil sloping up from a top of the wall or a building above the top of the wall. A surcharge increases very considerably the active earth pressure on the wall. This increase can be removed completely by a relieving platform.

Surcharged wall
A retaining wall carrying a surcharge, such as an embankment or road, above its top.

Surface activity
The physical and chemical manifestations of the surface electrical charges of a very fine mineral. Depending on the intensity of these charges the mineral is said to have a high or low surface activity.

Surface of rupture
The surface along which relative movement takes place when a soil slip occurs.

Surface tension
The property of water that permits the surface molecules to carry a small tensile force.
See also **Capillarity**.

Surface waterbar
A purpose-made thin strip of copper, plastic, rubber or steel and having edge ribs or dumb-bells. They are embedded into the concrete on each side of a construction joint to stop the ingress of water from the soil.

Surveying reference point
See under **Permanent settlement point**.

SUSIEPHONE
This stands for Scottish Utilities Service Information for Excavators.
This is a telephone system first inaugurated in 1980 to ease the exchange of

information between those involved in digging holes in the ground and the utility companies. It substitutes one freefone call for the many enquiries which would otherwise be necessary for information on the whereabouts of underground mains, pipes and cables.

Suspension grout
See under **Grout**.

Suspensions
Finely divided solid matter dispersed in a liquid.

Swallow hole

A cavity which occurs in some chalk, limestone and calcareous soil areas by solution; eventually the overburden collapses into it.

Swamp gas (CH_4)
Methane gas.

Swamp soil
A highly organic fibrous soil of very high compressibility found in swamp regions.

Swedish foil sampler
See under **Delft sampler**.

Sweet water
Water free from impurities and thus fit for drinking.

Swelling index
See **Expansion index**.

Swelling pressure
The pressure exerted by a contained clay when it absorbs water. It can amount to considerably more than the pressure of the overlying soil.

Swelling soil
A soil which appreciably swells on the introduction of water and conversely appreciably decreases in volume on the removal of water.

Swelling test
A double oedometer test to ascertain the swelling characteristics of a soil.

Swelling test apparatus
See **Geonor swelling test apparatus**.

Swinger
A pointed iron bar about 1 m long, used as a lever for moving runners.

Swivel rod
A boring rod with a swivelling eye at its upper end which permits a string of rods to rotate.

Symbols
See **Soil legend**.

T

Table
See **Water table.**

Tailings
See under **Colliery spoil.**

Tamper
See **Frog rammer.**

Tamping roller
The American name for a sheepsfoot roller.

Tandem roller
A roller having rolls of equal diameter one behind the other on the same track.

Tangent modulus
See **Initial tangent modulus.**

Tape extensometer
A portable instrument for measuring surface movement in adits, bridges, retaining walls, shafts and tunnels etc.
The instrument measures displacement between pairs of identical ball reference studs grouted into the structure. The tape extensometer unit comprises a stainless steel measuring tape, the free end of which is attached to a spring-loaded connector and thrust bearing assembly which locates on a ball reference stud. The fixed end of the tape unit is fitted to the tape reel housed in a light-weight body which has a ball location assembly identical to that on the free end of the tape.
See also **Magnetic probe** and **Potentiometric extensometers.**

Tapered pile
A pile that decreases in size with depth.
See also **Raymond pile.**

Taylor's curves
Curves produced by D W Taylor in 1948 to facilitate the working out of slope stability problems. The curves make use of a stability number which is non-dimensional.

T or **Tee beam footing**
A particular type of combined or cantilever footing where the individual column footings are joined by an inverted Tee beam.

TBM
Temporary bench mark.

Teeni slide rule
A slide ruler, adapted after M Teeni, for the rapid determination of soil densities.

Telescoping tube settlement gauge
See **USBR settlement gauge.**

Television camera
See **Borehole camera.**

Telltale

(1) A small, thin glass plate stuck to each side of a crack, or a small patch of cement mortar placed across a crack, to check for further movement of a structure. Further movement would thus crack the glass or mortar patch.

(2) A rod or pipe placed within the empty shell of a driven pile case and extending from the base to a reference point at the surface ground level to check for heave.
See also **Reference studs**.

Tension crack

A crack appearing some distance back from the top of a soil slope and parallel to it. The bottom of this crack, through which the slip circle will pass, is at a depth of $2C/\gamma\ (\pi/4 + \phi/2)$ below the surface,
where C is the soil cohesion,
γ is the soil density
and ϕ is the angle of shearing resistance of the soil.

Tension pile

A pile which is designed to resist an upward force.

Teredo

A marine organism which attacks timber piles in brackish or salt water.

Termite (Isoptera)

An insect which can eat its way through timber piles if they have not been treated with a chemical such as creosote. They are mostly found in tropical and sub-tropical regions.

Terrace

A naturally formed level flat or platform on a valley side. Terraces usually mark a pause in the down-cutting of a valley by the river, being remnants of old flooded plains. They are often overlain by spreads of gravel, hence gravel terrace or terraced gravels.

Terram

See **Fabric**.

Terra-probe

A method of compacting underwater sand fills or deep natural deposits of loose sand below the water table. It is similar in principal to the vibroflotation method except that no water jetting is needed.

Terzaghi, Karl *(1883–1963)*

The founder of soil mechanics, being the first person to make a comprehensive study of the engineering properties of soil.

Terzaghi analogy

Terzaghi's mechanical consolidation model consists of a vessel provided with a perforated piston, which is supported on a series of helical springs within the vessel.

In analogy with soil, the springs function according to the theory of elasticity and represent the soil particles and the volume of the vessel represents the voids of a soil with a specific porosity. If the volume of the vessel below the piston is filled with water, then by analogy, this means that all the soil voids are fully

saturated. The perforations in the piston represent the capillary passages in the soil.
At the instant when an external load is applied to the piston, the load is completely transferred to the water in the vessel, hence the water is stressed but the springs are not. In time the water will escape through the piston perforations, the escape velocity depending on the size and number of perforations, until equilibrium is obtained and the springs again support the external load.

Test anchor
A ground anchor specifically constructed at the start of a project on which a pull out test is made, to verify the suitability of the anchor design and construction for the particular site.

Test excavation
A small-scale soil excavation at or before the start of a project to check the field characteristics of both the soil and excavating equipment.

Test pile
A pile to which a load is applied to determine the load settlement characteristics of the pile and the surrounding ground.

Test pit
See **Trial pit**.

Textural classification system
See **Feret triangle**.

Thaw bowl
The profile of the thawed soil under a building originally constructed on frozen soil. It is due to a poorly heat insulated ground floor of the building.

Thaw coefficient
A coefficient to estimate the amount of thermal settling of frozen soil and the amount by which the permafrost will sag upon thawing.
When the thaw coefficient is greater than 0.02, permafrost soils are said to be sagging soils.

Thaw depth
The depth of complete melting of the ice present in soils of the annual frost layer overlying the permafrost.

Thawed soil
Soil which has previously been frozen.

Thermo-osmosis
The natural migration of water from the warmer parts of a soil layer towards the colder parts.

Thermopile
A tubular steel pile with a bottom steel plate cap resting on wooden planking on a gravel fill.
Cooling liquid gas condenses on the part of the pile above ground, which is made in the form of a finned radiator for better cooling, with a valve for filling the pile with the liquid.
The thermopile is suitable for use in permafrost soils.

Thickness gauge
A flat steel plate having a range of elongated holes of various dimensions, through which flaked soil can be passed to find its thickness.

Thin wall cut-off
See **Cut off wall.**

Thin wall sampler
See **Liner** and **Shelby tube sampler.**

Thixotropic fluids
Clays which are thixotropic show this by a weakening when they are remoulded and an increase in strength when they are allowed to stand undisturbed.

Thixotropy
The property possessed by some gels of becoming fluid when stirred and of returning to the jelly state when stirring ceases.

Three-dimensional consolidation
Settlement of an anisotropic soil where its horizontal permeability is much greater than its vertical permeability. Also, the settlement of a foundation which is small in plan area compared to the thickness of the consolidating stratum. See also **Consolidation** and **One-dimensional consolidation.**

Three in one cell
A triaxial compression cell for testing three samples, one by one, inside.

Three layer mineral
A mineral, such as montmorillonite, in which a single octahedral sheet is sandwiched between two tetrahedral sheets to give a 2:1 lattice structure.

Three way split former
See **Split former.**

Thrust borer
A piece of equipment for forcing a hole through soil, for instance to form a pedestrian subway under an existing road.

Thrust line
See **Line of thrust.**

Tidal lag
The delay between high tide (or low tide) in an estuary and the highest (or lowest) resulting level of the neighbouring ground-water.

Tidework
Soil investigation, or construction work carried out between high and low tide. According to the location of the work, the sea bed may or may not be exposed at low water.

Tieback
See **Ground anchor.**

Tie rod
A steel rod or bolt sometimes used instead of lacings between successive frames to take their mass and prevent movement of 'timbers'.

Till
Glacial drift of variable particle size; sometimes known as boulder clay.
See also **Raisin cake structure**.

Tiller rod
See under **Rod tiller**.

Tiltdozer
A blade similar to the bulldozer, fixed at right angles to the machine tracks so that it cannot be angled. The blade can be tilted up or down from the horizontal, thus giving a maximum difference in level between the two ends of 600 mm.
See also **Angledozer**, **Bulldozer** and **Back rippers**.

Tiltmeter
See **Inclinometer**.

Timber crib cofferdam
A framework of heavy timbers, either square or round in section and laced together in criss-cross fashion to form pockets up to 3.5 m square. The crib is then floated into position, suitably weighted with rock for sinking and then completely filled with rock, including smaller material to check infiltration.
See also under **Cofferdam**.

Timber grillage
See **Crib**.

Timbering
The support of the ground in excavations whether with wood, steel or concrete or light alloy posts, but usually with wood.

Timber pile
A pile made from hardwood which is resistant to water.

Time factor (T_v)
A dimensionless quantity equal to $c_v t/d^2$ used in one dimension consolidation analysis,
where c_v is the coefficient of consolidation,
t is the elapsed time,
and d is the length of the shortest drainage path.

Time lag
See **Tidal lag**.

Titration
See **Volumetric analysis**.

Toe
The bottom of a soil slope. The front of the base of a retaining wall.

Toe filter
A graded filter on the free side of an earth dam at its lower end, designed to protect it against piping.
See also **Earth dam**.

Tonne
A unit of mass equal to 1000 kilogrammes.

Tool
See under **Breaker tools.**

Tool lifter
See **Lifting dog.**

Top cap
See **Loading caps.**

Top frame
See **Ground frame.**

Top layer
See under **Pavement.**

Topography
The nature of the earth's surface as influenced by the hill and valley configurations.
See also **Ground roughness.**

Top-soil
The superficial skin of the deposits forming the earth's crust that has, by processes of weathering and the action of organic and other agencies, been transformed into material capable of supporting plant growth. The term thus embraces the upper or humus-bearing horizons of the soil of pedology.
See also **Aeration zone.**

Top spit
Loam containing much organic matter. It can carry little load and should therefore be removed in order to reach undisturbed soil below.

Torvane
A small hand held instrument for the rapid approximate determination of cohesive soil shear strength.

Total pressure (σ)
The pressure on a horizontal plane of soil due to the mass of the material above it plus any superimposed loads.
See also **Effective pressure.**

Total settlement
The total anticipated settlement of a structure.

Total stress (σ)
See **Total pressure.**

Total suction
The sum of the matrix and solute suctions.
See also **pF scale.**

Toughness index

The ratio $\dfrac{\text{index of plasticity}}{\text{flow index}}$

Tour
A work shift, usually of eight hours.

Towed vibrating roller
See **Vibrating roller**.

Track
A term used in the building industry for a trench for wall foundations.

Transducer
A piece of equipment which normally converts mechanical energy into electrical energy.
See also **Displacement transducer** and **Pressure transducer**.

Transient displacement gauge
An apparatus for the measurement of dynamic strain and long-term movement of soil beneath road pavements. The gauge comprises a casing fixed in the road wearing course, a sharp pointed rod anchored at its lower end at the elevation where the movement is to be measured and a displacement transducer to measure the movement of the datum rod relative to the casing.

Transported soil
Any soil which no longer covers the rock from which it was originally derived.

Transverse extensometer
An instrument for measuring precisely the profile of earth retaining structures. It consists of a vertical tensioned wire fixed between two anchor plates which makes contact with resistance elements. The resistance elements move with the adjacent structure, the wire being held by the end anchors. The horizontal movement of each resistance element is measured with respect to the steel wire by a Wheatstone bridge circuit.

Trapezoidal footing
A combined footing having the plan shape of a trapezoid. The location of the resultant of the column loads establishes the position of the centroid of the trapezoid.

Travel mixer
A self-propelled soil mixer which takes in soil at its front end and then discharges it after mixing it to the required water content and stabiliser content.

Travel time
A term used to describe the time taken by a scraper to transport its soil and return empty, all at a relatively constant speed. Some of the factors affecting travel time are altitude, coefficient of traction, distance of haul, rolling resistance, slope resistance, temperature and travel speed.
See also **Cycle Time**.

Tremie
A tube through which concrete is deposited under water.

Trench
An excavation whose length greatly exceeds its width and which has vertical sides capable of being supported by strutting from side to side, or battered sides requiring no support.
A trench is considered shallow if it does not exceed 1.5 m in depth; as medium between 1.5 and 4.5 m and as deep if it exceeds 4.5 m in depth.

Trench brace
See **Bracing**.

Trench cutting machine
A trench excavating machine which works in a similar manner to a chain cutter, ejecting the soil mechanically. The machine usually runs on rails and straddles the excavated trench.
See also **Roc-saw trencher**.

Trench fill
Completely filling a narrow excavated trench with load-bearing concrete, usually unreinforced, to form the foundation of a lightly loaded structure, such as the brick wall of a house.

Trench hoe
The name given in British practice to a special type of back acter used for excavating trenches.

Trial anchor
See **Test anchor**.

Trial excavation
See **Test excavation**.

Trial hole
See **Trial pit**.

Trial pit
A pit or shaft dug to inspect and determine *in situ* the type of soil below ground level.

Triangular classification diagram
See **Feret triangle**.

Triaxial cell
A cell for applying all-round pressure to a soil sample. Usually manufactured from aluminium alloy and perspex. It has an internal ram for applying vertical axial load to the sample and valves for (1) cell pressure, (2) pore water pressure and (3) pore water drainage.

Triaxial compression test
A compression test of a soil sample contained in a thin rubber sheath surrounded by liquid under pressure. The deformations, loads and pressures are recorded. Undrained tests are rapid, but drained shear tests are often called slow tests. Slow tests allow the soil to consolidate so that pore pressure does not develop.
See also **Consolidated undrained**, **Drained** and **Undrained tests**.

Trimming
The final tidying up of an earthwork surface.

Triple rod extensometer
An apparatus to monitor long-term movement in soil. It consists of three steel rods, separately attached to three anchor zones and passing through guide tubes to measuring points at the surface where the movement of a point on a

rod is observed relative to a fixed reference point. The apparatus is set up in a borehole which can be vertical, inclined or horizontal.

Triple tube core barrel
A double tube barrel with a longitudinally split third tube nested within the inner barrel. When the inner barrel is removed from the borehole, the split tube is pushed out to reveal an essentially undisturbed sample core.

Tripod derrick
A compact portable unit which assists in the removal of augers from the soil. It can be quickly assembled and when dismantled fits into the boot of a car.

Trisodium phosphate (Na_3PO_4)
A chemical used as a dispersing agent.

TRRL
Transport and Road Research Laboratory.

True cohesion (c_e)
See under **Angle of internal friction.**

Tube-a-manchette
See under **Sleeve grouting.**

Tube pile
See **Armco, Franki, Monotube** and **Steel piles.**

Tube sample
An undisturbed sample of soil obtained in a tubular soil sampler pushed into the soil at the bottom of a borehole or trial pit. After the soil sampler is removed the two ends of the tube are waxed and sealed and the whole is transported to a soils laboratory for testing.

Tube stop
See **Stop end tube.**

Tucking board
A narrow timber used behind walings in tucking frames.

Tucking frame
A frame in which the walings support the poling boards at their ends.

Tuff
A fine particled wind or water laid soil composed of small fragments ejected from erupting volcanoes.

Tunnel
A passage cut through a hill or mountain, or under a river.

Turfing
The practice of laying and pegging strips of grass turf on the slopes of newly constructed earthworks.

Turtle stone
See **Claystone.**

TV camera
See **Television camera.**

Twin shaft pier
See **Double shaft pier.**

Twin tube piezometer tip
See under **Piezometer tips.**

Two layer mineral
A mineral, such as kaolinite; in which a single tetrahedral sheet is joined to a single octahedral sheet to form a 1:1 lattice structure.

Two-shot grout
See under **Grout.**

Two-stage wellpoint system
See **Multistage well point system.**

Two-way drainage
See under **Drainage path.**

Two-way split former
See **Split former.**

Tyred roller
See **Pneumatic tyred roller.**

U

U1½ tube
A 38 mm (1½″) diameter tube for obtaining undisturbed samples of soil, either from site or from soil extruded from a U4 tube.

U4 tube
A 100 mm (4″) diameter tube for obtaining undisturbed samples of soil from site.

U abutment
An abutment where the wing walls are at right angles to the bridge pier.

Ultimate bearing capacity
See **Ultimate bearing pressure.**

Ultimate bearing pressure (q_f)
The pressure at which shear failure occurs in a soil. In plate bearing tests the ultimate bearing pressure can be taken as that pressure all over the plate at which the settlement amounts to one fifth of the plate width.

Ultimate pile load
The total load that a test pile can support.

Ultrasonic extractor
See **Pile extractor.**

Uncased concrete pile
See **Bored pile.**

Unconfined aquifier
Water in the pores of a soil below the free water surface or ground-water level. It is bounded on its upper side by air, in the pores of the soil, under atmospheric pressure.
See also **Confined water** and **Semi-confined water**.

Unconfined compression test
A compression test without lateral restraint to determine the shear strength of cohesive soil which is taken from shallow boreholes or trial pits during the design of roads, runways and light building construction.
At present the test is carried out on a soil sample 76 mm long by 38.1 mm (1 ½″) diameter in a portable unconfined compression testing machine. The test is done at the site as soon as possible after the extraction of the soil so that the soil moisture content does not have a chance to change. See also **Autographic unconfined compression apparatus.**

Unconfined compression test apparatus
See **Autographic unconfined compression apparatus.**

Unconfined well
See under **Well.**

Underflow
Movement of water in the soil, under ice, or under a structure.

Underpinning
The operation of providing a new foundation under parts of an existing foundation which has failed, or which is required to carry increased loading.

Under-reaming
The widening out of the foot of a bored hole or of a foundation pier to increase its area or to give an anchorage against lifting by wind or in permafrost regions. See also **Belling bucket.**

Undisturbed sample
A soil sample of cohesive soil from a borehole or trial pit which has been remoulded so little that it can be used for laboratory measurements of its strength without serious errors. Generally a soil sampler is driven into the ground, turned through 360° to shear the core at the foot of the tube, extracted from the borehole with the sample within it, filled with paraffin wax, covered with screw caps and sent to the laboratory for testing. Cores may also be tested at the site using the unconfined compression apparatus.

Undrained test
A shear test of a soil sample and usually carried out in a triaxial cell.
No drainage of the soil sample, and hence no dissipation of pore pressure, is allowed during the slow application of the all-round stress, neither is drainage allowed during the application of the deviator stress to the sample.
The test can be carried out on undisturbed samples of clay, peat and silt as a measure of the strength of the natural ground; and on remoulded samples of clay to measure sensitivity.
See also **Consolidated-undrained test.**

Unfrozen soil
Soil which has never been frozen, frosted, chilled or thawed.

Unified soil classification
This classifies soil on the basis of the texture and liquid limit. The system comprises fifteen soil groups, each identified by a two-letter symbol, the first symbol representing the type of soil and the second symbol indicating the plasticity of the soil.

Uniform flow
The flow of a liquid, the velocity vector of which is identical in magnitude for any given instant at every point.

Uniformity coefficient
See **Coefficient of uniformity.**

Uniform soil
A soil (usually sand), most of whose particles are of uniform size. It has a steep grading curve, that is, it is not a graded soil, ie D_{60}/D_{10} nearly equals 1.

Unit weight (γ)
The weight of a soil (or other material) divided by the corresponding volume.

Unsaturated flow
See **Non-saturated flow.**

Unstable condition
A term used in tunnelling, where the materials or soil being worked require support to hold them in position.
See also **Stable condition.**

Unsuitable fill
Material unsuitable to be used as fill. It includes:—
(1) Material from bogs, marshes and swamps.
(2) Logs, peat, perishable material and tree stumps.
(3) Material susceptible to spontaneous combustion.
(4) Material in a frozen condition.
(5) Clay having a liquid limit exceeding 80 and/or a plasticity index exceeding 55.
(6) Other materials having a high moisture content.

Unsymmetrical footing
A footing in which the resultant load does not coincide with the centre of gravity of the area of the footing.

Upconing
The phenomenon of a rising interface of ground-water due to a well above.

Uplift
See **Heave.**

Uplift pile
See **Tension pile.**

Upper soil belt
See under **Aeration zone.**

USBR
United States Bureau of Reclamation.

USBR settlement gauge
An instrument for the measurement and control of vertical movement in earth fill.
The equipment consists of telescoping steel tubes embedded in the earth fill, with each of the smaller diameter tubes anchored to the fill by a steel cross-arm. To take readings a portable measuring head is fixed to the upper tube and a torpedo suspended from a steel measuring tape is lowered down the tubing. Each time the torpedo passes the base of one of the smaller diameter tubes the tape is pulled taut. Spring-loaded pawls in the torpedo press outwards and engage the base of the smaller tube. Depths are measured from a reference point in the measuring head.

USDA
United States Department of Agriculture.

V

Vacuum sampling tubes
Special soil sampling tubes, which use the hydrostatic pressure at great ocean depths to drive the tube into the bottom sediments and/or to force the material into the sampling tube.

Vacuum testing of sand
A method of triaxial testing of a sand sample by maintaining a partial vacuum in the rubber membrane containing the sample, the outside of the tube being under atmospheric pressure. The resultant pressure on the tube is thus equal to the difference between the two pressures.

Vacuum well point
A well point system which maintains a vacuum in the well point and surrounding filter. It is used to assist the extraction of liquid from soil of low permeability.

Vadose water
The American term for 'held water'.

Valley spring
A spring which occurs where the water table rises above the actual ground surface formed by the V-shaped dip of a valley.
See also **Fault spring, Spring** and **Stratum spring**.

Van der Waal forces
Forces of attraction caused by the field generated by the spinning electrons surrounding the atomic nucleus. Such forces are large at spacings of the same order as the size of the atoms, but decrease inversely as the fifth power of the interparticle distance in the case of the very finest clays.
See also **Ionic bond**.

Vane borer
An *in situ* vane apparatus designed by the Royal Swedish Geotechnical Institute. It is designed for penetration boring and can be used without casing down to a depth of 30 m for determining clay soil shear strengths up to 100 kN/m^2.

The lower part of the device is forced into the clay with the crucifix vane withdrawn into a vane-shaped protection shoe. When the desired depth is reached the vane is unlocked and pushed down into the lower position and rotated at a constant rate of strain and the maximum torque is measured on an instrument at the ground surface.

Vane test
A four-bladed crucifix vane is inserted into the soil at the bottom of a borehole. It is rotated by a rod at the surface with a measured torque until the soil shears. A measurement of the *in situ* shear strength of the soil is thus obtained. It has been found to give consistent results down to 30 m depth, except for stiff-fissured clays and granular soils.
See also **Diamond vane, Laboratory vane test** and **Torvane.**

Vapour barrier
See **Damp proofing.**

Variable head permeameter
A falling head permeameter.
See also under **Permeameter.**

Varved clay
Clay formed in glacial lakes, which, owing to seasonal variations in the volume of the streams entering the lakes, ie, due to summer melting of ice, assumes a finely laminated structure, mud being deposited in winter and silt or fine sand in summer. The silty layers act as partings and enable the material to be split up by hand into thin laminae, hence the local names leaf clay and book clay.

V or **Vee drains**
See **Auxiliary drains.**

Ventilation modulus
The ratio of the area of the ventilation openings in the foundation walls of a building to the total area of the building floor.
The modulus is used to preserve the frozen state of the soil under a building.
See also **Thaw bowl** and **Thaw coefficient.**

Vermiculite
One of the clay minerals which has a structure similar to that of montmorillonite, except that the interstitial cations are principally magnesium, embedded in a thin layer of regularly arranged molecules of water.

Vertical sand drain
A boring through a clay or silty soil filled with sand or gravel to enable the soil to drain more easily.
Its special purpose is to accelerate the consolidation of a loaded cohesive soil.
See also **Drainage blanket** and **Sand wick.**

Vertical shore
See **Dead shore.**

Vertical tube settlement gauge
See **USBR settlement gauge.**

Very high liquid limit soil
A soil is said to have a very high liquid limit when this value lies between 70% and 90%.
See also **Liquid limit.**

Very high plasticity soil
A soil having a plasticity index greater than forty.
See also **Plasticity index.**

Very stiff
A clay is said to be very stiff when its undrained shear strength is greater than 150 kN/m^2.
See also under **Clay.**

Vibrated soil
See **Vibroflot.**

Vibrating extractor
See **Pile extractor.**

Vibrating hammer compaction test
A compaction test devised by the TRRL whereby the use of a standard falling rammer is replaced with a vibrating hammer.
See also **Proctor compaction test.**

Vibrating plate compactor
This consists of a heavy steel baseplate on which is mounted a vibrating mechanism. Soil compaction is achieved by a combination of vibration and pressure tamping.

Vibrating roller
A self-propelled or towed roller which is mechanically vibrated.

Vibrating wire extensometer
See under **Potentiometric extensometer.**

Vibrating wire load cell
The load cell contains two electromagnets which are positioned about 1 mm clear of a thin tensioned wire. One magnet excites the wire and as this oscillates the second magnet picks up the frequency of the vibrations, which can be recorded at a remote read-out location. The change in frequency of vibration of the wire is a measure of the strain, and hence the load, on the cell.

Vibrocompaction
See **Vibroflot.**

Vibro-concrete pile
A steel tube with a loose fitting cast-iron shoe is driven to a required set with a vibro piling hammer and concrete is poured into the tube. Extraction of the tube and formation of the concrete pile are effected by upward and downward blows of the hammer. Cage steel reinforcement is introduced if required.

Vibrodisplacement
A technique which involves a vibroflot penetrating the ground, the hole produced being infilled in stages with coarse backfill, the vibroflot compacting the material at each stage.

Vibroflot (or **vibroflotation**)
A geotechnical process for compacting clean sands and gravels. A vibrating cylinder, 450 mm in diameter and 2 m long, with water jets at both ends, is lowered into the ground by a crane. Its vibrations and the lower water jet allow it to sink easily and to compact the soil around it, and the upper jet enables it to be withdrawn easily. The compaction obtained can be seen at the surface by the diameter of the crater produced, often about 2.5 m. The sand must be clean. See also **Compozersystem, Sand piles** and **Terra probe.**

Vibroreplacement
A technique of forming a stone column in cohesive soil.
The method involves the displacement of cohesive soil radially from a vibration centre by a depth vibrator as it penetrates under its own mass and power of vibration. The cylindrical hole remaining on the withdrawal of the vibrator is infilled in stages with well-graded 75 mm to 10 mm angular stone and each stage is thoroughly compacted by reinsertion of the vibrator. The radial soil displacement at each stage continues until the forces of resistance are greater than the horizontal forces exerted by the vibrator.
The formation of the roughly cylindrical-shaped stone column ensures that the passive resistance of the surrounding cohesive soil is fully mobilised at very small horizontal radial strains when the stone column is subjected to vertical loading.

Vibro-tampers
Machines in which an engine-driven reciprocating mechanism acts on a spring system, through which oscillations are set up in a base-plate.

Vicksberg auger
A spoon type auger, hinged at its lower end to facilitate removal of the soil sample.

Vicksberg penetrometer
A hand held coned instrument for determining the penetration strength of soils.

Virgin branch
See **Virgin consolidation line.**

Virgin consolidation line
The nearly straight portion of a void ratio/pressure (log scale) consolidation curve for a remoulded soil.
See also **Compression index.**

Viscometer
An instrument for measuring viscosity.

Viscosity
A property of fluids, either liquid or gaseous, which is a measure of their resistance to flow.

Visual soil classification
See under **Rapid soil classification.**

Void measurement apparatus
A unit designed by the BRE to measure the void content of a sample of aggregate and or sand.

It comprises a sample container connected to a calibrated glass tube, the bottom of which is connected to a levelling bulb with water to a level indicated on the apparatus. The levelling bulb is then placed in its lower position and water is allowed to flow out until a pressure equilibrium exists in the system. This will directly relate to the amount of air forming the voids in the soil sample. The percentage of void content is obtained directly from the calibrated tube.

Void ratio (e)
The ratio of the volume of voids to the volume of the solids in a sample of soil or aggregate. For clays the void ratio must be defined at a stated pressure.
See also **Skeletal diagram**.

Voids
Spaces between separate soil particles and occupied by gas (air) or liquid (water or oil) or both.

Voigt model
See **Kelvin model**.

Volclay
A trade name for dry powdered bentonite.

Volume change apparatus
An apparatus to measure the change in volume of a soil sample in a drained test. This volume change will be caused by a change in cell pressure or in axial load being applied to the sample.
The volume can be measured in three ways based on (1) the volume of fluid entering the pressure cell to compensate for the change in volume of the sample, (2) the volume of water expelled from the voids of the soil, and (3) the direct accurate measurement of the change in length, and various diameters of the soil sample along its length.

Volume change modulus
See **Modulus of volume change**.

Volumeter
See **Volume change apparatus**.

W

Wakefield sheet pile
A simple wood sheet pile formed by nailing three planks together, with the centre plank offset 50 to 100 mm to form a tongue and a groove.

Wale
See **Waling**.

Waling
A horizontal beam of timber, steel or reinforced concrete, which supports piling boards, runners or sheet piles or the soil next to an excavation, and is held in place by horizontal struts.

Walking dragline
A dragline excavator which operates in the same way as the normal crawler tracked type. It differs only in the walking motion travelling device which replaces the crawler tracks on the base machine.

Wall anchor
See **Anchor block** and **Ground anchor**.

Wall drain
See **Back drain** and **Weephole**.

Wall footing
A continuous footing under a wall.

Wall friction
Friction between the back of a retaining wall and the retained soil. It generally improves the stability of the wall.
See also **Skin friction**.

Warp
Laminated sandy or silty clay formed by deposition in successive layers by periodic inundation of low-lying land, usually alluvial tracts.

Wash boring
A method of assisting an auger to drive a borehole with the aid of a jet of water.

WASHO
American Western Association of State Highway Officials.

Waste (or **spoil**)
The excess of excavation over fill.

Waste oil stabilisation
See under **Soil stabilisation**.

Waterbar
See **Surface waterbar**.

Water-bearing ground
Ground below the water table.

Water content (w)
See **Moisture content**.

Water displacement vessel
A steel open-topped cylindrical can of specific dimensions, with an overflow pipe situated approximately two thirds up. The apparatus is used in the determination of the dry density of soil by measuring its water displacement and mass.
See also **Pycnometer**.

Water glass injection
A method of soil stabilisation particularly suited to clean, relatively homogeneous sands having an effective size greater than 0.1 mm. The procedure consists of the successive injection of solutions of water-glass and calcium chloride, which react in the soil to form a cohesive binder.
See also **Soil stabilisation**.

Water jet
See **Pile water jet**.

Water level recorder
See **Dipmeter**.

Waterproof layer
See **Membrane**.

Water sampler
This comprises a rigid plastic cylinder of 500 ml internal volume, connected by nylon tubing to an air pump or compressed-air bottle, and a pressure gauge. To obtain a water sample an air pressure is applied to close a ball valve in the nose of the sampling cylinder. The pressure is then raised to a value greater than anticipated at the depth of water from which the sample is required. The cylinder, suspended on the nylon pressure tube, is lowered to the required depth when the air pressure in the cylinder is released; the ball valve opens and water enters the sampling cylinder. The ball valve is then closed by reapplying the initial air pressure, the sampling cylinder is brought to the surface and the water sample is poured into a sterile storage jar.

Water sensitive soil
See **Collapsible soil**.

Watershed
The summit and dividing line between two catchment areas from which water flows away in two directions.

Water soluble precondensates
Water soluble constituents of a condensation polymer in which the condensation process is largely arrested and which, when a suitable catalyst is added, is complete.
See also **Grouting**.

Water table
The level at which ground-water would stand in an unpumped borehole or well when equilibrium has been reached.

Weathering
Changes in colour, texture or chemical composition to the surface of a natural (or artificial) material due to the action of the weather.

Weather window
That part of the year when the weather is suitable for site investigation and for muck shifting, which cannot be carried out in adverse weather conditions.

Wedge
See **Folding-wedges** and **Page**.

Wedge theory
Coulomb's analysis of the force tending to overturn a retaining wall (1776). He based it on the mass of the wedge of soil which would slide forward if the wall failed.

Weephole
A hole to allow water to escape from behind a retaining wall and thus to reduce the pressure behind it.
See also **Back drain**.

Well
An artificial, usually vertical, cylindrical, perforated hydraulic structure, provided with a screen that taps ground-water from the voids of the soil which it penetrates. A well sunk into a water-bearing stratum and tapping free flowing water having a free ground-water table under atmospheric pressure is known by any of the following names: discharge, gravity, ordinary or unconfined.
A well sunk into and tapping water from an aquifier where the ground-water flows confined between two impermeable layers and under pressure greater than atmospheric is known by any of the following names: artesian, confined or pressure.

Well drain
Using a well to remove water from a stratum of soil by draining it into lower, pervious strata.

Well graded
A sample of soil is said to be well-graded when it contains a wide range of particle sizes and hence produces a smooth particle size distribution curve.

Well point
A 38 mm to 65 mm tube, sunk into a water-bearing soil, fitted with a close-mesh screen at the foot and connected through a header pipe to a suction pump at the top. Usually a number of well points are connected to one header. They are sunk outside an excavation to reduce the pumping required within it and to increase the strength of the ground by the flow of water towards the well points away from the excavation. An excavation in sand cannot become 'quick' if effectively wellpointed.
See also **Progressive system** and **Ring system**.

Well point system
See under **Well point**.

Westergaard theory
A method of rigid pavement design based on the plate loading test and which assumes that the loaded area is circular and that the pavement and subgrade are elastic.

West pile
See **Rotinoff pile**.

Wet analysis
The mechanical analysis of soil particles smaller than 0.06 mm diameter. It is done by mixing the sample in a measured volume of water and checking its density after various periods with a sensitive hydrometer. Several days may be needed for a test. Even with a centrifuge, to give the particles a settlement force of many hundreds of times gravity, the test is slow.

Wet density
See **Density** and **Partially saturated density**.

Wet pull
The suction caused by pulling a soil sampler out of the soil below the water table.

Wet sand process
A soil stabilisation process for sand. Special road oil and hydrated lime are mixed with the sand in the proportions of about 6% SRO and 2% lime.

Wetting
The process of watering soil to increase its moisture content and thus maintaining satisfactory working conditions.

Wheatstone bridge circuit
One of the simplest and best known bridge networks for measuring electrical resistances.

Wick drain
See **Sand wick**.

Wilson curves
Curves devised by Wilson to determine the centre of rotation of a soil slip circle.

Windblown deposit
Soil such as silt and sand which has been deposited under the action of wind.

Window
See **Weather window**.

Windrow
A ridge of soil consisting of the spill from a grader.

Wing wall (or **abutment wall**)
A wall at the abutment of a bridge, which extends beyond the bridge to retain the soil behind the abutment.
See also **Abutment**, **Spill through abutment**, **Stub wings** and **U abutment**.

Winkler ellipse
An ellipse drawn with one-half of the major axis equal to the major principal stress σ_1 and one-half of its minor principal stress σ_3.
The drawn ellipse can be used to determine the magnitude and direction of any stress as the radial distance to the ellipse when that stress is acting on any plane passing through the same point of application as the principal stresses σ_1 and σ_3.

Wire cutter
See **Wire saw**.

Wire line drilling
See **Double tube core barrel**.

Wire saw
A saw consisting of a thin wire stretched between the ends of a U-frame and used for trimming soil samples.

Wolmar Fellenius
See **Fellenius, Wolmar**.

Wooden pile
See **Timber pile**.

Working chamber
The chamber at the bottom of a pneumatic caisson, in which workmen can operate.

Working load
The load which a pile is designed to carry.

Working pile
One of the piles forming the foundation of a structure.

Worm auger
See **Helical auger**.

X

XY/t Recorder
See **Potentiometric recorder**.

Y

Yield point
The minimum unit stress at which a cohesive soil (and other structural materials) will deform without an increase in the load is called the yield point.

Yield stress model
A simple rheological model to represent a perfect strain lack response of a material.
See also **Rheological model**.

Yield value
The shear strength which has to be exceeded for slurry flow to commence.
Yield value is dependent upon the surface properties, density of solids and chemical nature of the slurry. It is measured by means of a rotational viscometer.

Young's modulus
This indicates the manner in which a soil (or other material) can resist elastic deformation. It is the ratio of stress to resultant strain.

Z

Zero air voids
The condition of a soil in which the voids are just completely full of water.

Zero air voids line
See **Saturation line**.

Zero isochrome
See under **Isochrome**.

Zone
Four arbitrary zones of grading into which sand and fine aggregate can be placed according to their distribution of particle sizes.

Zone of saturation
See **Ground-water**.